中等职业学校1+X证书实训指导

用微课学·网络设备互连与配置
（华为版）

主 编 黄 磊 门雅范
副主编 张 鑫 连 静 高 玲

电子工业出版社
Publishing House of Electronics Industry
北京·BEIJING

内 容 简 介

本书是以校企双元培养高技能人才为编写背景，依据企业岗位人才需求，结合教学规律编写的项目式实用教程。本书以职位岗位技能建设要求为主线来组织内容，通过教材的学习，学生可以从零基础入门，迅速成长为具备网络设计和搭建能力的熟手或高手。

本书通俗易懂、图文并茂、过程严谨，使学生和教师在轻松愉快的氛围中学习和教授网络设备互连与配置的相关知识和实践技能。同时，本书在每个项目中都配有思考与实训，并穿插许多小技巧和知识补给，有助学生提高学习效率和扩展知识结构。

未经许可，不得以任何方式复制或抄袭本书之部分或全部内容。

版权所有，侵权必究。

图书在版编目（CIP）数据

用微课学·网络设备互连与配置：华为版 / 黄磊，门雅范主编．—北京：电子工业出版社，2022.10

ISBN 978-7-121-44414-2

Ⅰ．①用… Ⅱ．①黄… ②门… Ⅲ．①网络设备—配置—中等专业学校—教材 Ⅳ．①TN915.05

中国版本图书馆 CIP 数据核字（2022）第 188148 号

责任编辑：关雅莉　　文字编辑：张　慧
印　　刷：三河市华成印务有限公司
装　　订：三河市华成印务有限公司
出版发行：电子工业出版社
　　　　　北京市海淀区万寿路 173 信箱　邮编　100036
开　　本：880×1 230　1/16　印张：12　字数：276.48 千字
版　　次：2022 年 10 月第 1 版
印　　次：2022 年 10 月第 1 次印刷
定　　价：42.80 元

凡所购买电子工业出版社图书有缺损问题，请向购买书店调换。若书店售缺，请与本社发行部联系，联系及邮购电话：（010）88254888，88258888。

质量投诉请发邮件至 zlts@phei.com.cn，盗版侵权举报请发邮件至 dbqq@phei.com.cn。

本书咨询联系方式：（010）88254550，zhengxy@phei.com.cn。

前　言

本书由职业学校专业教师与华为公司技术人员共同编写。本书为响应《国家职业教育改革实施方案》要求，以促进产教融合、校企"双元"育人、培养高素质技能人才为目标，以就业为导向，突出"校企共培"的人才培养模式特征。本书按照教育部颁布的《中等职业学校计算机网络技术专业标准（试行）》编写，尊重中等职业学生的年龄特点和认知规律，适合企业用人需要。本书在编写过程中，力求突出以下特色。

1．分层导入。本书按照从一层介质互连、二层设备至三层设备等由浅入深、逐层递进的方式编写，既符合网络体系结构，又符合中等职业学生认知规律。

2．案例真实。本书采用的任务案例均来源于企业真实的工作场景，体现理论和实践相结合的特点。同时，本书在编写过程中以"适用、实用、够用"为原则，将理论知识与实际应用相结合，避开空洞的理论讲解及无理论支撑的实践操作，符合企业用人需求。

3．注重实训。本书以应用为核心，以培养学生实际动手能力为重点，采用真实的项目背景和任务导入要求，将操作技能融入每一个任务案例中，使学生通过动手完成每一个任务案例，强化操作训练，从而达到掌握知识与提高技能的目的。本书在编写过程中，力求做到"做、学、教"的统一。

4．结构合理。本书紧密结合职业教育和学生学习的特点，借鉴近年来职业教育课程改革和教材建设的成功经验，由浅入深、由静到动，先设备后协议，从静态路由到动态优化，培养学生从新手转变为熟手。

5．实用性强。本书在知识讲解与任务实施的基础上，根据理论难度提供"知识补给""小技巧"和"任务拓展"等内容，便于教师教学和学生自学。

6．资源丰富。本书配备包括教学PPT、配置文档、教学素材、操作视频等内容在内的教学资源包，为教师备课和学生自学提供了资源平台。

本书共分为五个项目，项目一主要介绍OSI体系中一层的物理介质，包括制作互连介质、互连PC与交换机、互连交换机与交换机、互连路由器与路由器、测试网络连通性等内容；项目二主要介绍二层设备交换机的安装与调试，包括交换机的外观与接口、交换机的管理方法、隔离业务流量、VLAN之间的通信技术、生成树技术等内容；项目三主要介绍三层设备——三层交换机和路由器，包括配置三层交换机的路由功能、安装与连接路由器的方法、认识路由器的指示灯与接口、配置单臂路由、静态路由等方法；项目四主要介绍网络优化，包括配置动态路由协议、维护与管理中型局域网、配置与优化RIP路由协议、配置单区域OSPF路由协议等内容；项目五主要介绍无线与安全设备，包括认识无线AC及无线AP、规划无线局域网、安装无线AP、实现无线局域网配置、测试无线局域网的性能及覆盖范围、了解安全设备防火墙等内容。

本书教学总课时为86学时，项目一、项目二、项目三为必修内容，项目四、项目五为选修内容，教师可根据学生的接受能力与专业需求，灵活进行内容选择与课时调配。下表为课时分配参考表。

项目	课程内容	课时分配		
		讲授	实训	小计
项目一	网络设备互连	4	4	8
项目二	接入二层设备	10	10	20
项目三	三层业务互访	8	10	18
项目四	动态管理路由	10	10	20
项目五	无线与安全设备	8	12	20
合　计		40	46	86

本书由黄磊、门雅范担任主编，张鑫、连静、高玲担任副主编，陈向末、张娇、寇艳艳、程雅青、王永锋参加了本书的编写，全书由蒋明辉进行审校。在此向参与本书编写工作的各位教师致以衷心的感谢。

由于时间仓促，加之计算机网络技术突飞猛进，虽然我们力求做到精益求精，但难免有所疏漏，敬请读者批评指正。

编者

目 录

项目 1 网络设备互连 ... 1

- 1.1 PC 接入交换机 ... 2
  - 1.1.1 非屏蔽双绞线和屏蔽双绞线 ... 7
  - 1.1.2 非屏蔽双绞线（UTP）的两种类型 ... 8
  - 1.1.3 DTE 和 DCE ... 8
- 1.2 交换机间的光纤互连 ... 9
  - 1.2.1 初识光纤 ... 15
  - 1.2.2 光纤类型 ... 16
  - 1.2.3 光模块 ... 16
  - 1.2.4 光纤连接器 ... 17
- 1.3 路由器的广域网互连 ... 17
- 思考与实训 1 ... 22

项目 2 接入二层设备 ... 23

- 2.1 认识交换机 ... 24
  - 2.1.1 交换机控制线缆连接 ... 27
  - 2.1.2 交换机硬件接口 ... 28
  - 2.1.3 交换机指示灯说明 ... 29
- 2.2 配置交换机 ... 31
  - 2.2.1 常用的命令行视图 ... 34
  - 2.2.2 退出命令行视图 ... 35
  - 2.2.3 帮助功能和命令自动补全功能 ... 35
- 2.3 隔离部门网络 ... 36
  - 2.3.1 VLAN 简介 ... 40
  - 2.3.2 VLAN 划分 ... 41
  - 2.3.3 VLAN 操作命令 ... 41
- 2.4 相同 VLAN 中的通信 ... 42
  - 2.4.1 VLAN 主要的优点 ... 47
  - 2.4.2 以太网接口类型 ... 48
  - 2.4.3 恢复接口上 VLAN 的默认配置 ... 48

2.5 避免形成网络环路（STP 技术）·· 49
　　2.5.1 STP 的定义和目的 ··· 57
　　2.5.2 STP 的收敛过程 ··· 57
　　2.5.3 常见 STP 操作 ·· 58
思考与实训 2 ··· 59

项目 3　三层业务互访 ·· 61
3.1 利用三层交换机实现部门网络互通 ·· 62
3.2 安装路由器 ··· 68
　　3.2.1 HUAWEIAR2220E 路由器的接口与槽位 ································· 72
　　3.2.2 HUAWEIAR2220E 路由器的指示灯 ···································· 72
3.3 管理路由器 ··· 75
3.4 实现不同业务网络互访 ·· 82
3.5 升级路由器 ··· 87
3.6 访问外部网络 ··· 97
思考与实训 3 ·· 110

项目 4　动态管理路由 ··· 112
4.1 认识动态路由协议 RIP ·· 113
4.2 认识动态路由协议 OSPF（单区域）······································· 121
　　4.2.1 OSPF 的报文 ··· 130
　　4.2.2 OSPF 支持的网络类型 ··· 131
4.3 转换网络地址 ·· 132
　　4.3.1 网络地址转换 NAT ·· 135
　　4.3.2 ACL 访问控制列表 ·· 136
4.4 动态分配 IP 地址 ·· 136
4.5 分发静态路由协议 ··· 141
思考与实训 4 ·· 146

项目 5　无线与安全设备 ··· 150
5.1 认识无线设备 ·· 151
5.2 规划 WLAN ··· 159
5.3 配置 AP 上线 ··· 161
5.4 测试 WLAN 性能 ··· 169
5.5 认识防火墙 ·· 173
5.6 划分安全区域 ·· 179
思考与实训 5 ·· 184

项目 1

网络设备互连

☆ 项目背景

近日,宇信公司搬入政府规划的企业园区办公,需要部署全新的企业网络,目前已经完成网络设备及通信线缆的采购与验收。新入职的助理工程师小李,是刚从某职业院校计算机网络专业毕业的新人,缺少项目实践经验。在项目经理高工程师的指导下,他开始学习制作各种网络传输介质,以便连接网络设备。

1.1 PC接入交换机

➢ 任务情景

为搭建办公室网络,小李需要制作2根双绞线网线,用于连接办公用PC和交换机,要求网线制作工艺符合规范并且美观,设备间连通测试正常。

➢ 任务分析

- ➢ 了解双绞线(UTP)的基本结构和特性;
- ➢ 掌握制作合格T568A/T568B双绞线网线的方法;
- ➢ 掌握测试UTP网线连通性的方法。

➢ 实施准备

1. HUAWEIS5720-28P-EI-AC交换机1台;
2. PC 1台;
3. 超5类非屏蔽双绞线1卷;
4. RJ45水晶头4个;
5. 压线钳1把;
6. 网线测试仪1台。

➢ 实施步骤

1. 依据办公室的实际情况,选择合理的布线路径,用卷尺测量出需要连接的2个网口之间的路径距离为2m,连接交换机与PC的双绞线网线类型为直通线,要求遵循国际线序标准。双绞线网线线序标准如表1-1-1所示。

表1-1-1 双绞线网线线序标准

类型	线序
T568A	白绿—绿—白橙—蓝—白蓝—橙—白棕—棕
T568B	白橙—橙—白绿—蓝—白蓝—绿—白棕—棕

🔥 小贴士

如果布线路径经过活动连接部件(如布线从可开闭的箱门连接到箱体内部)时,则网线裁剪长度应留有能够保证活动连接部件流畅运动的余量。

2. 准备工具和材料。根据实施准备的要求准备网线测试仪、压线钳、水晶头等网线制作工具和材料,如图1-1-1所示。

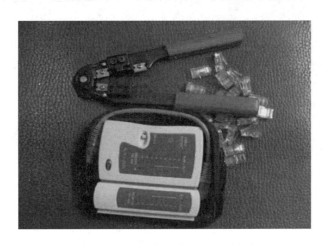

图 1-1-1　网线制作工具和材料

3．观察双绞线的结构。小李打开 1 卷超 5 类非屏蔽双绞线，认真观察双绞线的结构特征。双绞线内部有 4 对 8 芯绝缘铜芯导线，每 2 根相互缠绕，铜芯导线的绝缘层分别涂有不同的颜色，其结构如图 1-1-2 所示。

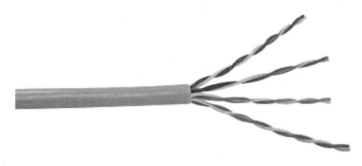

图 1-1-2　双绞线的结构

4．剪断双绞线。利用压线钳的剪裁刀口，将双绞线剪裁为需要的长度。

5．双绞线剥皮。利用压线钳的剪线刀口，将线头剪齐，再将线头放入专用的剥线刀口，稍微用力握紧压线钳，慢慢旋转，用刀口划开双绞线的保护胶皮，如图 1-1-3 所示。

图 1-1-3　双绞线剥皮

小贴士

剥线时，应避免剥线过长或过短，剥线长度以 2~3cm 为宜，不可太用力，否则容易把网线的线芯剪断。

6．铜芯线排序。首先把铜芯线按 T568B 的标准线序拆开、理顺、捋直，然后按照 T568B 标准线序排列，即按照白橙-橙-白绿-蓝-白蓝-绿-白棕-棕的顺序排列。用双手抓住排列好的铜芯线两端，反向用力，并上下拉扯，在此过程中应尽量保持铅芯线扁平排列，如图 1-1-4 所示。

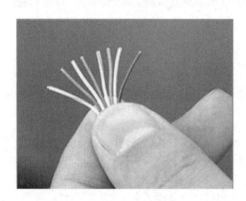

图 1-1-4　铜芯线排序

7．剪齐铜芯线。利用压线钳的剪裁刀口，把排列好的铜芯线顶部裁剪整齐，长度差控制在 1mm 范围内，如图 1-1-5 所示。

图 1-1-5　剪齐铜芯线

小贴士

剪裁时，应使铜芯线垂直插入剪裁刀口下方，使露在保护层外的铜芯线长度为 1.5cm 左右。

8．插入水晶头。右手手指掐住铜芯线，左手拿水晶头，使水晶头的弹簧片朝上，双手缓缓用力，将 8 根铜芯线同时延水晶头内的 8 个线槽插入水晶头，一直插入线槽的顶端。从水晶头的顶部观察，确保每根铜芯线都紧紧地顶在水晶头的末端，同时应确保线序正确，如图 1-1-6 所示。

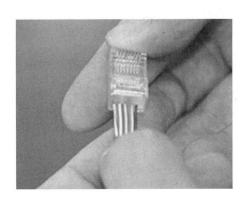

图 1-1-6　插入水晶头

小贴士

确保铜芯线的外皮在水晶头内，否则会造成水晶头松动。

9. 压制水晶头。把水晶头完全插入压线钳，用力握紧压线钳，直至听到"咔嚓"声，说明压制成功，可重复压制多次，如图 1-1-7 所示。

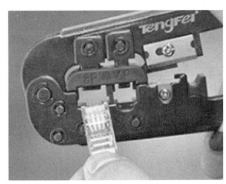 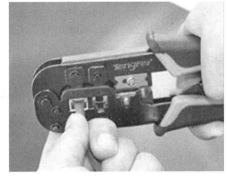

图 1-1-7　压制水晶头

10. 压制双绞线的另一端。按照步骤 5～9 压制双绞线的另一端，如图 1-1-8 所示。

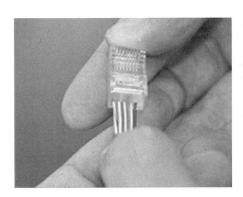 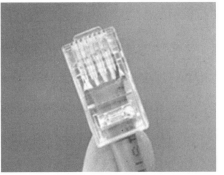

图 1-1-8　压制双绞线的另一端

11. 网线测试。将做好的网线的两头分别插入网线测试仪，并打开开关，如果两侧的指示灯同步亮起，则表示网线制作成功，如图1-1-9所示。

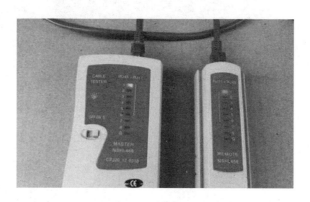

图1-1-9　网线测试

12．设备间互连测试。按照以上制作网线的步骤，制作2根网线，用来连接HUAWEIS5720-28P-EI-AC交换机与PC，从而实现网络设备与终端设备间的互连。交换机间互连使用交叉型双绞线，连接方式如图1-1-10所示。

本任务采用制作的直通型双绞线（UTP）连接交换机与PC，连接方式如图1-1-11所示。连接完成后，测试并观察设备接口指示灯状态，HUAWEIS5720-28P-EI-AC交换机接口绿灯长亮，PC终端网卡RJ-45接口绿灯常亮、橙灯闪烁表示正常，指示灯状态如图1-1-12所示。

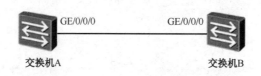

图1-1-10　交换机间的连接方式

图1-1-11　交换机与PC间的连接方式

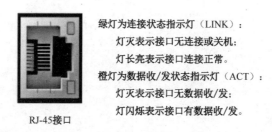

图1-1-12　PC终端网卡RJ-45接口指示灯状态

小贴士

直通型双绞线网线两端的线序相同，同采用T568A标准或T568B标准，用来连接不同类型设备。交叉型双绞线两端的线序不同，一端采用 T568A 标准，另一端采用 T568B 标准，用于连接同类型设备。

任务总结与思考

本任务重点讲述物理介质双绞线的制作与使用。在实践过程中，应重点掌握直通型双绞线的制作流程，并能制作合格的双绞线。

思考以下两个问题。

1. 如果制作交叉型双绞线，那么在直通型双绞线的基础上应怎么调整线序？
2. 根据制作双绞线的实际操作流程和测试结果，总结制作双绞线的注意事项。

知识补给

1.1.1 非屏蔽双绞线和屏蔽双绞线

双绞线可分为非屏蔽双绞线（Unshielded Twisted Pair，UTP），如图 1-1-13 所示，以及屏蔽双绞线（Shielded Twisted Pair，STP），如图 1-1-14 所示。屏蔽双绞线的特点是在双绞线与外层绝缘封套之间有一个金属屏蔽层，可以屏蔽电磁干扰。

图 1-1-13 非屏蔽双绞线（UTP）

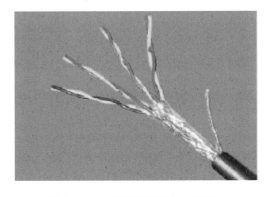

图 1-1-14 屏蔽双绞线（STP）

目前，UTP 因其价格低廉，制作和连接方便，在网络工程中被广泛使用，本书只介绍 UTP

的制作与连接方法。

1.1.2 非屏蔽双绞线（UTP）的两种类型

1. 标准网线，通常称为直通网线，用于连接不同类型设备，如路由器与交换机互连、交换机与 PC 互连。

2. 级联网线，通常称为交叉网线，用于连接同类型设备，如路由器局域网口互连、交换机之间互连、两台 PC 通过网卡口互连，也包括连接路由器局域网接口与 PC 端网卡口的直连。

3. 网线的线序要求。

交叉网线：一端采用 T568A 标准，另一端采用 T568B 标准，如图 1-1-15 所示。

直通网线：两端同采用 T568A 标准或 T568B 标准，通常采用 T568B 标准，如图 1-1-16 所示。

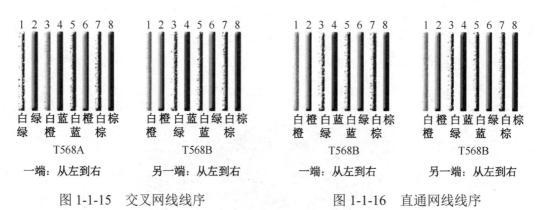

图 1-1-15　交叉网线线序　　　　图 1-1-16　直通网线线序

小贴士

目前，大部分厂商提供的设备均内置 MDIX 功能，设备网口为自适应网线类型，因此无须区分交叉网线和直通网线。

1.1.3 DTE 和 DCE

1. DTE 是数据终端设备，如终端，是广义的概念。通常，广域网的常用 DTE 包括路由器、终端主机。

2. DCE 是数据通信设备，如 MODEM 等，是用于连接 DTE 的通信设备。通常，广域网的常用 DCE 包括 HUB、广域网交换机、MODEM。

3. 相同类型设备互连，即 DTE 与 DTE、DCE 与 DCE 互连，用交叉线，如 PC-PC、PC-路由器、路由器-路由器、HUB-交换机、交换机-交换机、HUB-HUB 互连等。

4. 不同类型设备互连，即 DCE 与 DTE 用直通线互连，如 PC-HUB、PC-交换机、路由器-HUB、路由器-交换机互连等。

任务拓展

机房共有 40 台 PC，需要接入 2 台交换机。现需要制作一根交叉型 UTP，用于连接 2 台交换机。请根据本任务中的 UTP 制作流程，制作一根 1 米长的 UTP，并进行测试。

小技巧

1. 剥线时，不可太深，也不可太用力，否则容易把双绞线剪断。
2. 一定要把每根铜芯线都捋直、排列整齐，捋线时，不要太用力，以免将线拗断。
3. 把 8 根铜芯线都捋直后，剪齐，使露在保护层外皮的双绞线长度为 1.5cm 左右。
4. 把铜芯线插入水晶头时，8 个线头都要紧紧地顶到水晶头的末端。
5. 水晶头应压住双绞线外皮，以保证双绞线不从水晶头中脱出。

1.2　交换机间的光纤互连

➤ 任务情景

办公室网络搭建完成后，需要把办公室局域网连接到公司核心交换机，由于互连距离超出 100m，且带宽需要满足 1GE/10GE 要求，因此必须使用光纤跳线将办公室交换机 A 连接到核心交换机 B。小李需要在交换机上安装光模块，并制作 LC 型光纤跳线。在安装光模块及进行互连时，既需要符合技术规范又需要美观，且要保证设备间的连通测试正常。

➤ 任务分析

- ➢ 学会安装 HUAWEIS5720-28P-LI-AC 交换机光模块；
- ➢ 学会制作 LC 型光纤跳线；
- ➢ 学会测试设备间光链路的连通性。

➤ 实施准备

1. HUAWEIS5720-28P-LI-AC 交换机 2 台；
2. GE 光模块 2 个；
3. 2m 单模光纤 2 根；
4. 光纤制作工具 1 套（米勒钳、光纤切割刀）。

➢ 实施步骤

1. 认识交换机接口。打开HUAWEIS5720-28P-LI-AC交换机包装箱，观察其外观及接口情况，如图1-2-1所示。

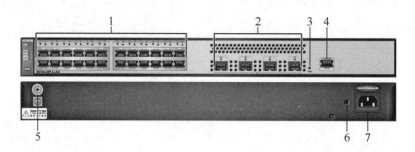

图 1-2-1　HUAWEIS5720-28P-LI-AC

通过观察接口标识及设备说明书，小李了解到该交换机包括24个双绞线接口、4个光纤接口，接口说明如表1-2-1所示。

表 1-2-1　交换机 HUAWEIS5720-28P-LI-AC 接口说明

序号	接口名称	功能说明
1	24 个 10/100/1000BASE-T 以太网电接口	使用 5 类以上双绞线
2	4 个 1000BASE-X 以太网光纤接口	支持 GE 光模块，带宽支持 10Mbit/s/100Mbit/s/1000Mbit/s
3	1 个 PNP 按钮	预留按钮，当前不支持使用
4	1 个 Console 接口	控制台接口
5	接地螺钉	配套使用接地线缆
6	交流端子防脱扣插孔	为安装交流端子防脱扣预留的插孔，交流端子防脱扣不随设备发货
7	交流电源插座	配套使用交流电源线缆

【提示：接口 1000BASE-T 中的"T"表示双绞线介质，1000BASE-X 中的"X"表示光纤介质。】

2. 选择交换机使用的光模块和光连接器类型。

按照设备说明书要求，此交换机的1000BASE-X 以太网光纤接口支持 GE 光模块（见图1-2-2）进行 10Mbit/s/100Mbit/s/100Mbit/s 带宽的接收和发送，光连接器支持 LC/PC 型单模光纤，如图1-2-3所示，光纤接头如图1-2-4所示。

图 1-2-2　GE 光模块

图 1-2-3　LC/PC 型单模光纤　　　　图 1-2-4　LC/PC 型光纤接头

【提示：LC 是指光纤接头类型，"LC/" 后的 "PC" 等是指光纤接头截面类型。PC 在电信运营商的设备中应用最广泛，其接头截面是平的。】

3．认识光纤制作工具。光纤制作工具有米勒钳和光纤切割刀，如图 1-2-5 所示。

图 1-2-5　光纤制作工具

4．把光纤尾帽套入光纤，剥去 5cm 的光纤护套（外绝缘皮），用米勒钳剥去剩余部分的纤芯涂层，如图 1-2-6～图 1-2-8 所示。

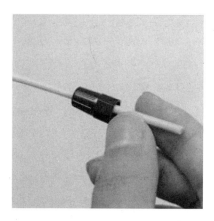 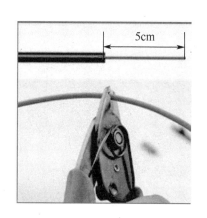

图 1-2-6　套入光纤尾帽　　　　图 1-2-7　剥去光纤护套

5．将光连接器的锁扣移至底部，光纤从光连接器尾部穿入孔中，顶到限位处，并确保光纤尾部略有弯曲，如图 1-2-9 所示。插入时提紧外绝缘皮，防止皮线松动影响通光效果。

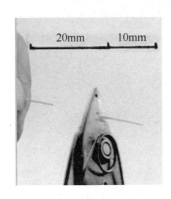 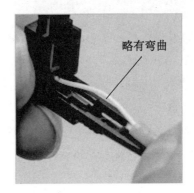

图 1-2-8 剥去纤芯涂层　　　　　图 1-2-9 纤芯插入光连接器

6. 保持光纤对接状态，上移锁扣直至锁紧，盖上盖子，拧紧光纤尾帽，如图 1-2-10 和图 1-2-11 所示。

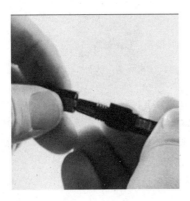 

图 1-2-10 盖上盖子　　　　　图 1-2-11 拧紧光纤尾帽

7. 套上光连接器的外套并旋紧。注意，外套的光面向上、凸起面向下，如图 1-2-12 所示。完成操作后，用激光笔检验通光效果，如图 1-2-13 所示。

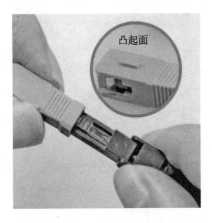 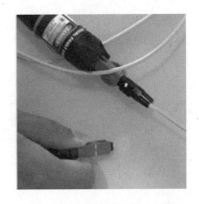

图 1-2-12 套上外套　　　　　图 1-2-13 检验通光效果

【提示：安装时施工人员应保持手部清洁干燥，切割固定长度的光纤时，请正确使用定长器或光纤切割刀。】

8. 重复步骤 4～7，安装光纤的另一端 LC 连接器及另一根 LC 光纤。

9. 拔出交换机 A 的 GE0/0/1 接口的防尘塞，将光模块安装到光接口，如图 1-2-14 所示。

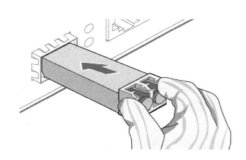

图 1-2-14　安装光模块

【提示：安装光模块时，如果按一个方向无法完全插入，则勿强行推入，可将光模块翻转 180°后重新插入，光模块簧片会发出"啪"的响声，此时如果光模块不能被拔出，则说明光模块已正确安装到位；如果光模块能够被拔出，则说明光模块安装不正确，需重新安装。】

10. 去掉做过导通性测试并贴有标签的光纤上的防尘塞，将光纤插入接口，如图 1-2-15 和图 1-2-16 所示。

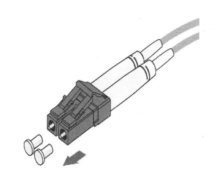

图 1-2-15　去掉防尘塞

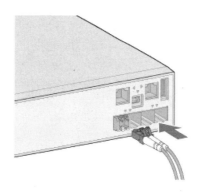

图 1-2-16　将光纤插入接口

11. 用一对光纤线连接交换机的 GE0/0/1 接口，如图 1-2-17 所示，SWA 的 GE0/0/1 的 TX 接口连接 SWB 的 GE0/0/1 的 RX 接口，SWB 的 GE0/0/1 的 TX 接口连接 SWA 的 GE0/0/1 的 RX 接口。

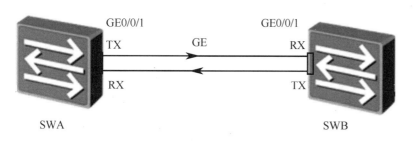

图 1-2-17　交换机光纤连接示意图

交换机间光纤互连

观察该 GE 光模块指示灯是否均绿灯长亮。通过命令行查询该 GE 光模块的状态可知，物理层和链路层均为 UP 状态。

12．验证交换机 A 的接口状态。

```
[SWA]dislay interface GigabitEthernet 0/0/1   //查询GigabitEthernet 0/0/1接口状态
GigabitEthernet0/0/1 current state : UP       //物理层当前状态为UP
Line protocol current state : UP              //链路层当前状态为UP
```

13．验证交换机 B 的接口状态。

```
[SWB] dislay interface GigabitEthernet 0/0/1  //查询GigabitEthernet 0/0/1接口状态
GigabitEthernet0/0/1 current state : UP       //物理层当前状态为UP
Line protocol current state : UP              //链路层当前状态为UP
```

14．光模块互通测试。

确定交换机光模块物理连接正常之后，为两台交换机配置 VLAN1 三层接口地址。其中，配置交换机 A 的 VLAN1 三层接口地址为 10.0.0.1/24，配置交换机 B 的 VLAN1 三层接口地址为 10.0.0.2/24。

配置交换机 A 的 VLAN1 三层接口地址。

```
[SWA]interface vlanif 1                //进入VLAN1三层接口
[SWA-Vlanif1]ip address 10.0.0.1 24    //配置VLAN1三层接口地址
```

配置交换机 B 的 VLAN1 三层接口地址。

```
[SWB]interface vlanif 1                //进入VLAN1三层接口
[SWB-Vlanif1]ip address 10.0.0.2 24    //配置VLAN1三层接口地址
```

15．分别在两端测试交换机连通性。对两台交换机进行 ping，ping 结果应正常。

```
[SWA]ping 10.0.0.2     //交换机A ping交换机B的VLAN1三层接口地址
PING 10.0.0.2: 56  data bytes, press CTRL_C to break
    Reply from 10.0.0.2: bytes=56 Sequence=1 ttl=254 time=9 ms
    Reply from 10.0.0.2: bytes=56 Sequence=2 ttl=254 time=2 ms
    Reply from 10.0.0.2: bytes=56 Sequence=3 ttl=254 time=2 ms
    Reply from 10.0.0.2: bytes=56 Sequence=4 ttl=254 time=2 ms
    Reply from 10.0.0.2: bytes=56 Sequence=5 ttl=254 time=2 ms

  --- 10.0.0.2 ping statistics ---
    5 packet(s) transmitted
    5 packet(s) received
    0.00% packet loss
    round-trip min/avg/max = 2/3/9 ms
```

【提示：SWA 通过 ping 命令向 IP 为 10.10.10.2 的 SWB 发送 5 个 ICMP 包，收到了 5 个答复（Reply），每个 ICMP 包的字节数（bytes）均为 56，序列号（Sequence）分别为 1～5，生存时间（ttl）均为 254，耗费时间（time）除了第一个包，其余均为 2ms，丢包率为 0%，说明二者之间的网络已经互通，状态良好。】

```
[SWB]ping 10.0.0.1      //交换机B ping交换机A的VLAN1三层接口IP地址
 PING 10.0.0.1: 56  data bytes, press CTRL_C to break
    Reply from 10.0.0.1: bytes=56 Sequence=1 ttl=255 time=32 ms
    Reply from 10.0.0.1: bytes=56 Sequence=2 ttl=254 time=2 ms
    Reply from 10.0.0.1: bytes=56 Sequence=3 ttl=254 time=2 ms
    Reply from 10.0.0.1: bytes=56 Sequence=4 ttl=254 time=2 ms
    Reply from 10.0.0.1: bytes=56 Sequence=5 ttl=254 time=2 ms
```

```
--- 10.0.0.1 ping statistics ---
  5 packet(s) transmitted
  5 packet(s) received
  0.00% packet loss
  round-trip min/avg/max = 2/8/32 ms
```

【提示：SWB 通过 ping 命令向 IP 为 10.10.10.1 的 SWB 发送 5 个 ICMP 包，收到了 5 个答复（Reply），每个 ICMP 包的字节数（bytes）均为 56，序列号（Sequence）分别为 1～5，生存时间（ttl）为 255，耗费时间（time）除了第一个包，其余均为 2ms，丢包率为 0%，说明二者之间的网络已经互通，状态良好。】

 小贴士

1. 交换机的光模块支持热插拔，操作时应注意光模块要插拔到位。
2. 光纤较为脆弱，操作时应注意不要随意弯折。
3. 不使用光纤接头时应用防尘帽盖好，以免灰尘污染光纤接口造成误码率过高。

 任务总结与思考

本任务重点讲述光纤物理介质的原理和使用方法，在任务实施过程中，应重点掌握交换机光口互连通断测试，掌握测试流程。

思考以下两个问题。

1. 如果光纤的收/发接口连接反了，即 TX 接口对 TX 接口、RX 接口对 RX 接口，则会产生什么结果？

2. 根据交换机光模块通断测试的实际操作流程，总结操作过程中有哪些注意事项。

知识补给

1.2.1 初识光纤

光纤呈圆柱形，由纤芯、包层与涂层三个部分组成，如图 1-2-18 所示。

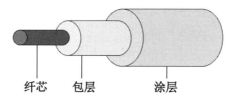

图 1-2-18 光纤的构造

1. 纤芯主要采用高纯度的二氧化硅（SiO_2）组成，并掺有少量的掺杂剂，以提高纤芯的光折射率。

2. 包层也是由高纯度的二氧化硅组成的，并掺杂一些掺杂剂，主要功能是降低包层的光折射率。

3. 涂层由丙烯酸酯、硅橡胶、尼龙组成，以增强机械强度和可弯曲性。

1.2.2 光纤类型

1. 单模光纤和多模光纤。

因为纤芯的粗细不同，所以光纤所支持的传输模式数量多少也不同。光纤按照传输模式分为单模光纤和多模光纤，如图 1-2-19 所示。

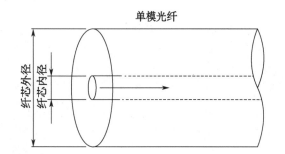
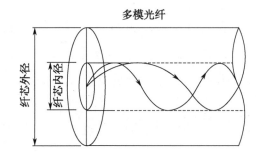

图 1-2-19　单模光纤和多模光纤

（1）当光纤纤芯的几何尺寸与光信号波长相差不多时，光纤只允许光以一种传播模式在其中传播，这样的光纤叫作单模光纤。单模光纤的纤芯直径较小，通常为 5～10μm。

（2）当光纤纤芯的几何尺寸远大于光波波长时，光在光纤中会以几十种或更多种传播模式进行传播，这样的光纤叫作多模光纤。多模光纤的纤芯直径较大，通常为 50μm 左右。

注意：从光纤的外观上来看，两种光纤的区别不大；带有塑料护套的光纤的直径都小于 1mm。

2. 通信波长。

适合通信的光纤工作波长（工作窗口）主要有 850nm、1310nm、1550nm。

其中，850nm 为短波信号，1310nm 和 1550nm 为长波信号。

3. 单模光纤、多模光纤与通信波长的关系。

（1）1310nm 和 1550nm 工作窗口用于单模传输。

（2）850nm 工作窗口只用于多模传输。

1.2.3　光模块

1. 光模块又称收发器，"光收发器=光发射器+光接收器"。光发射器包含一个激光器或发光二极管，以用于创建光脉冲，负责在发送方向将数字信号调制到光脉冲上，以完成电光转换。光接收器包含一个探测光的光学传感器，负责在接收方向将光信号上的数字信号解调出来，并还原为数字信号以完成光电转换。

2．光模块指标。

（1）平均发送光功率：在发送"0""1"码等概率调制的情况下，光发射器输出的光功率值，单位为dBm。

（2）光接收器灵敏度：在保证规定的误码率情况下，光接收器所需的最小光功率值，单位一般为dBm。

（3）光接收器过载光功率：在保证规定的误码率情况（如 BER＝1×10-10）下，光接收器所允许的最大光功率值，单位为dBm。

1.2.4 光纤连接器

1．如图1-2-20所示，常见光纤连接器种类如下。

（1）FC：圆头尾纤连接器。

（2）SC：方头尾纤连接器。

（3）LC：小型尾纤连接器。其体积小，集成度高，在新产品中被广泛使用。数据通信产品的光纤连接器大多为LC。

（4）MTRJ：收发一体光纤连接器。

图1-2-20 光纤连接器

2．光纤与光纤连接器的耦合精度非常高，出厂时的光纤已带有光纤连接器，不支持现场制作。光纤出厂时的长度一般根据前期勘测工程师现场勘测的光纤布放长度确定。余量光纤按照施工要求盘纤，盘纤直径最小为10cm，否则会产生弯曲损耗。

 小技巧

1．交换机光模块互连时需注意设备上侧光模块指示灯 ACT/LINK 的状态，以确定物理层通断状态。

2．交换机光模块互连时应注意收/发线不要接反，否则易造成物理连接不通。

1.3 路由器的广域网互连

➢ 任务情景

由于宇信公司企业网需融合传统语音网络，因此需要将路由器RTA使用广域网方式和其他路由器互连，而且互连时要使用支持语音业务的接口。小李决定使用 serial 接口（又称串行接口）进行互连。此时，小李需要连通链路、做好配置，然后实现业务的互通。

> 任务分析

> 认识 serial 接口的线缆；
> 掌握 serial 接口 IP 地址的配置方法；
> 验证 serial 接口的连通性；
> 查看 serial 接口的连接状态。

> 实施准备

1．HUAWEIAR2220-AC 路由器 2 台；
2．serial 线缆 2 根；
3．SA 单板 1 块。

> 实施步骤

1．HUAWEIAR2220-AC 路由器的后面板的槽位都是空的（见图 1-3-1），并没有放置业务单板。

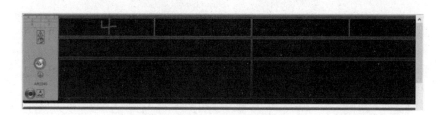

图 1-3-1　路由器后面板

2．选用 SA 单板，如图 1-3-2 所示。

图 1-3-2　SA 单板

3．把该单板放置在路由器的第 4 槽位，放置后的路由器面板如图 1-3-3 所示。

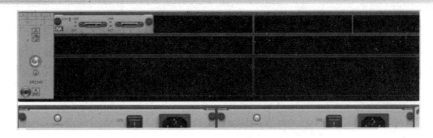

图 1-3-3　放置后的路由器面板

4. 使用 serial 线缆（见图 1-3-4）连接两台路由器的 serial 接口。

图 1-3-4　serial 线缆

5. 完成连接。网络拓扑图如图 1-3-5 所示。

图 1-3-5　网络拓扑图

路由器广域网互连实验

6. 启用 RTA 的 4/0/0 接口，配置 RTAserial4/0/0 接口地址，用来验证链路的连通性。

```
<Huawei>system-view//进入系统视图
[Huawei] sysname RouterA//修改路由器名称为RouterA
[RouterA] interface serial 4/0/0   //进入serial接口4/0/0
[RouterA-Serial4/0/0]link-protocol ppp//配置接口封装的链路层协议为PPP
[RouterA-Serial4/0/0] ip address 10.1.1.1 30//配置接口地址
[RouterA-Serial4/0/0] quit//返回上级模式
```

7. 启用 RTB 的 4/0/0 接口，配置 RTBserial4/0/0 接口地址，用来验证链路的连通性。

```
<Huawei>system-view//进入系统视图
[Huawei] sysname RouterB//修改路由器名称为RouterB
[RouterB] interface serial 4/0/0   //进入serial接口4/0/0
[RouterB-Serial4/0/0]link-protocol ppp//配置接口封装的链路层协议为PPP
[RouterB-Serial4/0/0] ip address 10.1.1.2 30//配置接口地址
[RouterB-Serial4/0/0] quit//返回上级模式
```

8. 验证配置结果。

```
# 查看接口的详细信息，可以看到接口的物理状态和链路层协议状态都是UP
[RouterA] display interface Serial 4/0/0   //查看S4/0/0接口信息
Serial4/0/0 current state : UP//UP表示该接口物理层已连通
Line protocol current state : UP//链路层协议UP表示该链路可以传递数据
```

```
Last line protocol up time : 2012-07-18 11:12:29
Description:HUAWEI, AR Series, Serial4/0/0 Interface
Route Port,The Maximum Transmit Unit is 1400, Hold timer is 10(sec)
Internet Address is 10.1.1.1/30
Link layer protocol is PPP
LCP opened, IPCP opened
Last physical up time   : 2008-01-09 12:25:52
Last physical down time : 2008-01-09 12:25:51
Current system time: 2008-01-09 19:18:44
Physical layer is synchronous, Virtualbaudrate is 72000 bps
Interface is DTE, Cable type is V35, Clock mode is DTECLK1
Last 300 seconds input rate 0 bytes/sec 0 bits/sec 0 packets/sec
Last 300 seconds output rate 0 bytes/sec 0 bits/sec 0 packets/sec
Input: 47266 packets, 662564 bytes
  Broadcast:              0,  Multicast:              0
  Errors:                 0,  Runts:                  0
  Giants:                 0,  CRC:                    0
  Alignments:             0,  Overruns:               0
  Dribbles:               0,  Aborts:                 0
  No Buffers:             0,  Frame Error:            0
Output: 47267 packets, 662592 bytes
  Total Error:            0,  Overruns:               0
  Collisions:             0,  Deferred:
DCD=UP DTR=UP DSR=UP RTS=UP CTS=UP
    Input bandwidth utilization  : 0.09%
Output bandwidth utilization : 0.09%

# 查看接口的路由表，以RouterA为例，可以看到存在到达对端的路由信息
[RouterA] display ip routing-table
Route Flags: R - relied, D - download to fib
------------------------------------------------------------------
Routing Tables: Public
        Destinations : 5       Routes : 5
Destination/Mask    Proto  Pre  Cost     Flags NextHop       Interface
     10.1.1.0/30    Direct  0    0         D   10.1.1.1       Serial4/0/0
     10.1.1.1/32    Direct  0    0         D   127.0.0.1      Serial4/0/0
     10.1.1.2/32    Direct  0    0         D   10.1.1.2       Serial4/0/0
     10.1.1.3/32    Direct  0    0         D   127.0.0.1      Serial4/0/0
    127.0.0.0/8     Direct  0    0         D   127.0.0.1      InLoopBack0
    127.0.0.1/32    Direct  0    0         D   127.0.0.1      InLoopBack0
```

【提示：路由表中存在该接口生成的直连路由表项，说明该链路已正常，可以用来传输用户的业务报文。】

```
# RouterA和RouterB可以互相ping通。以RouterA为例，ping RouterB，有如下结果
[RouterA] ping 10.1.1.2
  PING 10.1.1.2: 56  data bytes, press CTRL_C to break
    Reply from 10.1.1.2: bytes=56 Sequence=1 ttl=255 time=90 ms
    Reply from 10.1.1.2: bytes=56 Sequence=2 ttl=255 time=50 ms
    Reply from 10.1.1.2: bytes=56 Sequence=3 ttl=255 time=50 ms
    Reply from 10.1.1.2: bytes=56 Sequence=4 ttl=255 time=40 ms
    Reply from 10.1.1.2: bytes=56 Sequence=5 ttl=255 time=30 ms
  --- 10.1.1.2 ping statistics ---
```

```
    5 packet(s) transmitted
    5 packet(s) received
    0.00% packet loss
round-trip min/avg/max = 30/52/90 ms
```

【提示：PC1 通过 ping 命令向 IP 地址为 10.1.1.2 的 PC2 发送 5 个 ICMP 包，收到 5 个答复（Reply），每个 ICMP 包的字节数（bytes）均为 56，序列号（Sequence）分别为 1～5，生存时间（ttl）为 255，耗费时间（time）除了第一个包，其余均为 50ms，丢包率为 0%，说明二者之间网络互通，状态良好。】

任务总结与思考

serial 接口又称串行接口，用于路由器与路由器之间的互连，请思考以下两个问题。

1．串行接口可用于哪种网络场景？
2．串行接口可用于承载哪种业务？

知识补给

serial 接口是最常用的广域网接口之一，可以在同步方式或异步方式下工作，因此又称同异步串口。

SA 单板在同步方式下工作，企业可以通过同步串口接入传输网，实现企业分部与总部之间的数据传输，每个接口的带宽可达到 8.192Mbit/s。

• serial 接口可以在数据终端设备 DTE 和数据通信设备 DCE 两种方式下工作，在 serial 接口插入 DTE 线缆的设备称为 DTE，在 serial 接口插入 DCE 线缆的设备称为 DCE 设备。在一般情况下，DTE 可接收 DCE 提供的时钟。

• 链路层协议类型为 PPP（Point-to-Point Protocol），这是一种点到点的链路层协议，主要用于在全双工的同异步链路上进行点到点的数据传输。

• 支持的链路层协议包括 PPP、帧中继和 HDLC。

任务拓展

在选择 SA 线缆前，需首先确认对端设备的类型（对端的同/异步方式、DTE/DCE 方式等）、接入设备所要求的带宽，以及要承载的业务。

小贴士

SA 单板释义：S 表示 Sync，即同步串口；A 表示 Async，即异步串口。

思考与实训 1

一、填空题

1. T568A 标准的线序是_____。
2. T568B 标准的线序是_____。
3. 直通型双绞线的线序要求是_____，交叉型双绞线的线序要求是_____。
4. 光纤的构造包括_____、_____、_____。
5. 按照纤芯直径大小不同，光纤分为_____、_____。
6. HUAWEIS3700 交换机光模块的类型为_____（在 LC、SC、FC 中选择）。

二、判断题

1. PPP 是 serial 接口的一种封装协议。   （   ）
2. serial 接口是一种串行接口。   （   ）
3. 路由器上的 serial 接口板支持热插拔。   （   ）
4. 路由器支持两端接口地址不在同网段的互通。   （   ）
5. 路由器的 serial 接口板可以更换槽位。   （   ）

三、实训操作

1. 制作一根 2m 长的直通型双绞线和一根 1m 长的交叉型双绞线。
2. 制作一根单模 LC 型光纤，完成交换机间的光模块互连，并测试相关通断命令。

项目 2

接入二层设备

☆ 项目背景

宇信公司办公网络项目所需的网络设备选型及采购已经完成,已进入项目实施阶段,当前首先需要部署的是各部门的办公网络。小李在高工程师的指导下,开始进行二层设备的互连及调试工作,同时在实践中学习并掌握设备配置技巧。

2.1 认识交换机

➢ 任务情景

宇信公司的资产采购部门购买了一批华为HUAWEIS3700系列交换机,主要用于各部门办公网络的搭建。在开始搭建办公网络之前,小李需要首先了解这批交换机的特点和性能,以便登录交换机进行相应配置。通常,首次登录新的交换机时需通过Console接口登录并进行初始配置。

➢ 任务分析

- ➢ 认识交换机硬件接口及功能;
- ➢ 了解交换机指示灯状态信息;
- ➢ 掌握通过Console接口登录的方法。

➢ 实施准备

1. HUAWEIS3700-26C-HI交换机1台;
2. PC 1台;
3. Console通信线缆1根。

➢ 实施步骤

1. 将Console通信线缆的DB-9(孔)接头插入PC的串口(COM)中,再将RJ-45接头插入设备的Console接口中,如图2-1-1所示。

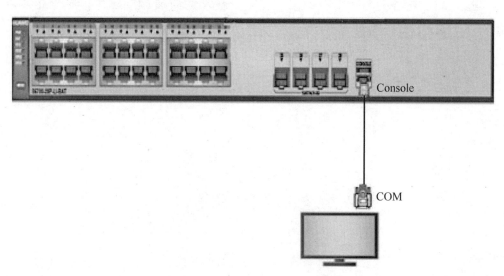

图2-1-1 通信线缆连接示意图

小贴士

如果用户使用的 PC 是便携式计算机或没有串口的计算机，则需要使用 USB 转串口的转接线。首先按照说明书安装转接线驱动软件，然后将转接线的 USB DB-9（孔）接头插入 PC 的 USB 接口中，再将 RJ-45 接头插入设备的 Console 端口中。

2．交换机接入电源，使用 Console 通信线缆连接 PC 的 COM 端口后，打开"计算机管理"窗口，如图 2-1-2 所示，在该窗口中确认 COM 端口的 ID。

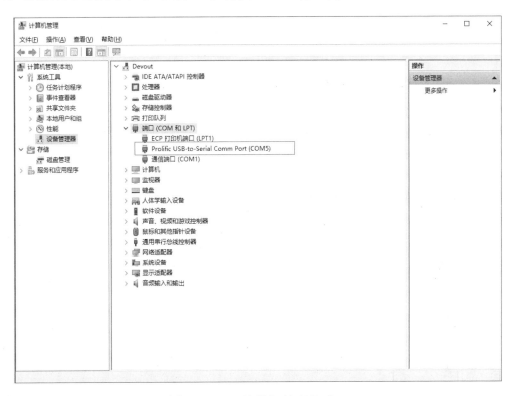

图 2-1-2 "计算机管理"窗口

3．在连接交换机的 PC 上打开 CRT 软件（SecureCRT 7.0）主页面，单击快速连接图标按钮，如图 2-1-3 所示。

图 2-1-3 CRT 软件主页面

4．在弹出的"快速连接"对话框中，按如图 2-1-4 所示配置端口、波特率、数据位、停止位等参数，完成后单击"连接"按钮。

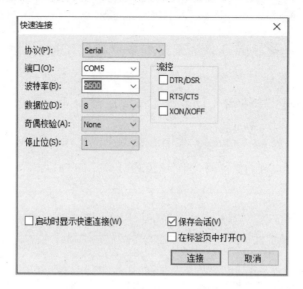

图 2-1-4　CRT 软件参数配置

🌱 小贴士

在默认情况下,"RTS/CTS"选项处于勾选状态,因此需要取消勾选,否则无法在终端页面中输入命令行。

5. 单击"连接"按钮后,终端页面会出现提示信息,以提示用户输入用户名和密码。首次登录时默认的用户名为 admin,默认密码为 admin@huawei.com,系统提示必须重新设置密码。(以下显示信息仅为示意。)

```
Login authentication
Username:admin
Password:          //输入默认密码admin@huawei.com
Warning: The default password poses security risks.
The password needs to be changed. Change now? [Y/N]: y
//需要更改密码,是否现在创建(是/否)? 是
#提示:首次登录交换机时需要更改默认密码
Please enter old password:   //请输入旧密码
#提示:输入默认密码admin@huawei.com
Please enter new password:    //输入新密码
Please confirm new password:    //再次输入新密码
The password has been changed successfully//密码修改成功
<HUAWEI>
```

🌱 小贴士

➢ 密码为字符串形式,区分大小写,长度范围为 8~16 位。密码至少包含两种类型的字符,包括大写字母、小写字母、数字及特殊字符。特殊字符不包括?和空格。

➢ 采用交互方式输入的密码不会在终端屏幕上显示出来。

➢ 用户设置密码成功后，如果用户没有修改验证方式及验证密码，那么当再次登录设备时，用户的验证密码即为初次登录时所设置的验证密码。此时用户可以输入命令，对交换机进行配置，如果需要帮助则可以随时输入？符号以唤起系统帮助。

6．检查配置结果。

配置完成后，用户使用 Console 用户页面重新登录交换机时，只有输入通过上述步骤设置的用户名和验证密码才能通过身份验证，成功登录交换机。

任务总结与思考

若要对一台新出厂的设备进行业务配置，则首先需要从本地登录设备。本任务重点讲述首次登录设备时通过 Console 接口登录的具体操作方法。

思考以下两个问题。

1．设备支持的首次登录方式除通过 Console 接口登录交换机外，还有没有其他的登录方式？
2．如果忘记了 Console 接口的登录密码，那么应如何进行密码恢复操作？

知识补给

2.1.1 交换机控制线缆连接

首次使用交换机设备时，一般使用产品随机附带的 Console 通信线缆进行连接，再利用 PC 端的终端软件登录设备进行配置，将 Console 通信线缆的 DB-9 接头插入 PC 的 9 芯（针）端口（端口标志是 COM）的插口中，再将 RJ-45 接头插入交换机的 Console 接口中。

通信线缆连接拓扑图如图 2-1-5 所示，通信线缆实物外观如图 2-1-6 所示，通信线缆结构图如图 2-1-7 所示，通过 Console 接口连接设备如图 2-1-8 所示。

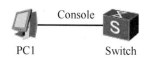

图 2-1-5　通信线缆连接拓扑图

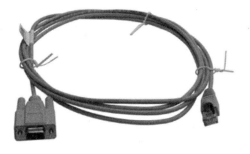

图 2-1-6　通信线缆实物外观

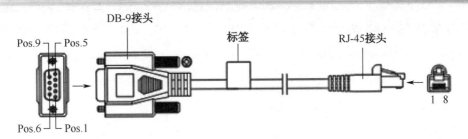

图 2-1-7　通信线缆结构图

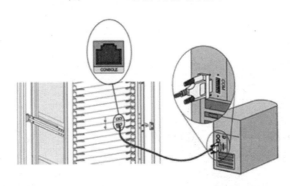

图 2-1-8　通过 Console 接口连接设备

2.1.2　交换机硬件接口

交换机硬件接口如图 2-1-9 所示。

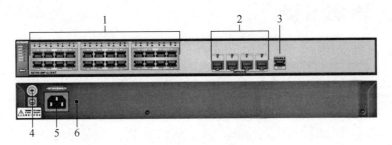

图 2-1-9　交换机硬件接口

1. 交换机硬件接口说明如表 2-1-1 所示。

表 2-1-1　交换机硬件接口说明

编号	接口	说明
1	24 个 10/100/1000BASE-T 以太网接口	主要用于十兆/百兆/千兆业务的接收和发送
2	4 个 1000BASE-X 以太网光纤接口	● 使用 GE 时不支持百兆传输，仅可用于千兆业务的接收和发送； ● 使用 GE 时支持十兆/百兆/千兆业务的接收和发送
3	1 个 Console 接口	用于连接控制台，实现现场配置功能
4	接地螺钉	配套使用接地线缆
5	交流电源插座	配套使用交流电源线缆
6	交流端子防脱扣插孔	为安装交流端子防脱扣预留的插孔，交流端子防脱扣不随设备发货

2. 交换机各硬件接口应连接线缆如图 2-1-10 所示。

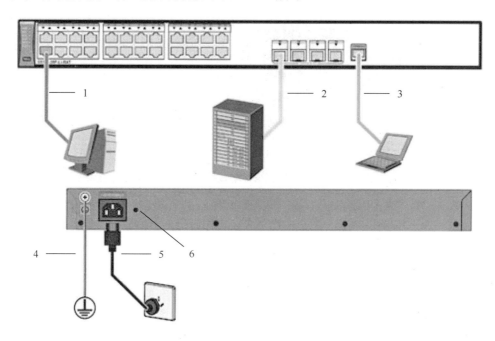

图 2-1-10　各硬件接口连接线缆

交换机硬件接口应连接线缆如表 2-1-2 所示。

表 2-1-2　交换机硬件接口应连接线缆

序号	名称	对端连接设备	备注
1	网线	交换机或计算机等	以太网电接口，用于数据业务的接收和发送
2	光纤	上层网络设备	安装前应检查光纤连接器是否被污染。若是，则建议用无尘棉布或擦纤盒擦拭光纤连接器
3	Console 通信线缆	维护终端（一般为计算机）	Console 接口，设备初次上电使用时需要通过其进行配置，以实现现场配置功能
4	接地线缆	M4 端连接到设备；M6 端连接到保护地	孔径较小的一端为 M4 端，孔径较大的一端为 M6 端
5	交流电源线缆	外部供电设备	使用当地制式的交流电源线缆
6	交流端子防脱扣插孔	—	为安装交流端子防脱扣预留的插孔，交流端子防脱扣不随交换机发货

2.1.3　交换机指示灯说明

交换机指示灯如图 2-1-11 所示。

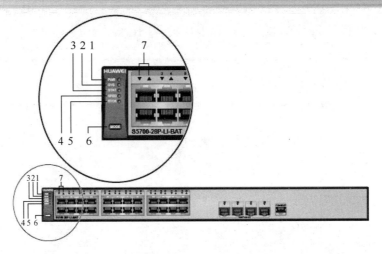

图 2-1-11 交换机指示灯

交换机各指示灯功能说明如表 2-1-3 所示。

表 2-1-3 交换机指示灯说明

编号	指示灯（按钮）	颜色	指示灯（按钮）状态及含义
1	电源指示灯：PWR	—	常灭：交换机未上电
		绿色	常亮：电源供电正常
		黄色	常亮：交换机处于备份电源供电状态，主电源故障
2	系统运行状态灯：SYS	—	常灭：系统未运行
		绿色	• 快闪：系统正在启动过程中 • 慢闪：系统正常运行中
		红色	常亮：交换机不能正常启动，或者运行中有温度、风扇异常告警
3	status 模式状态灯：STAT	绿色	• 常灭：表示没有选择 status 模式 • 常亮：表示业务端口指示灯为默认模式，默认模式下端口为 status 模式
4	speed 模式状态灯：SPED	绿色	• 常灭：表示没有选择 speed 模式 • 常亮：表示业务端口指示灯暂时用来指示端口的速率，45s 后自动恢复到默认模式
5	stack 模式状态灯：STCK	绿色	• 常灭：端口无连接或被关闭 • 常亮：端口有连接 • 闪烁：端口在发送或接收数据
6	模式切换按钮：MODE	—	• 按一次按钮则切换到 speed 模式，此时业务端口指示灯暂时用来指示各端口的速率状态 • 再按一次按钮则恢复默认状态
7	GE 电端口：端口从 1 开始编号，从下到上、从左到右顺序编号	绿色	• 常灭：端口无连接 • 闪烁：端口有连接

2.2 配置交换机

配置交换机

➢ **任务情景**

宇信公司为组建各部门办公局域网络，新购置了一批 HUAWEIS3700 系列交换机，用于实现终端设备的接入。在连接终端设备之前，小李需要首先熟悉交换机的各种配置模式及基本配置命令。

➢ **任务分析**

- ➢ 了解交换机的各种配置模式；
- ➢ 掌握交换机的基本配置命令。

➢ **实施准备**

1. HUAWEIS3700-26C-HI 交换机 1 台；
2. PC 1 台；
3. Console 通信线缆 1 根。

➢ **实施步骤**

1. 按照如图 2-2-1 所示拓扑图，使用 Console 通信线缆连接 PC 和交换机设备。

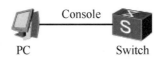

图 2-2-1 通信线缆连接拓扑图

2. 在 PC 上打开终端仿真软件，新建连接，设置连接的端口及通信参数，如图 2-2-2 所示。

图 2-2-2 参数设置

> 🌱 **小贴士**
>
> 因 PC 端可能存在多个连接端口，所以这里需要选择连接 Console 通信线缆的端口 ID。一般情况下，选择的端口是 COM1。

3．单击"确定"按钮后，按 Enter 键即可进入交换机的用户配置模式，终端页面会出现提示信息，提示用户输入用户名和密码。首次登录时默认的用户名为 admin，默认密码为 admin@huawei.com，系统提示必须重新设置密码。（以下显示信息仅为示意。）

```
Login authentication
Username:admin
Password:          //输入默认密码admin@huawei.com
Warning: The default password poses security risks.
The password needs to be changed. Change now? [Y/N]: y  //修改登录密码
Please enter old password:        //输入默认密码admin@huawei.com
Please enter new password:        //输入新密码
Please confirm new password:      //再次输入新密码
The password has been changed successfully//密码修改成功
<HUAWEI>
```

4．成功登录到交换机后，可以执行 display version 命令，查询交换机的软件版本与硬件信息。

```
<Huawei>display version //查询软件版本与硬件信息
Huawei Versatile Routing Platform Software
VRP (R) software, Version 5.110 (S3700 V200R001C00)
Copyright (c) 2000-2011 HUAWEI TECH CO. LTD
Quidway S3700-26C-HI Routing Switch uptime is 0 week, 0 day, 0 hour, 39 minutes
#提示：命令回显信息中包含了VRP版本、设备型号和启动更新时间等信息
```

5．修改系统时间。

VRP 系统会自动保存时间，但如果时间不正确，则可以在用户视图下执行 clock timezone 命令和 clock datetime 命令来修改系统时间。

```
<Huawei>clock timezone Beijing add 08:00:00//设置时钟时区
#提示：可以修改Beijing字段为当前地区的时区名称。如果当前时区位于UTC+0时区的西部，则需要把add字段修改为minus
<Huawei>clock datetime 12:00:00 2020-3-11//设置日期和时间
<Huawei>display clock//查询系统日期和时间
#提示：执行display clock命令可查看生效的新系统时间
2020-3-11 12:00:00
Friday
Time Zone(Local) : UTC+08:00
```

> 🌱 **小贴士**
>
> 地球分为 24 个时区，每个时区都有自己的本地时间。在国际无线电通信场景下，均使用一个统一的时间，称为通用协调时（Universal Time Coordinated，UTC）。北京时区是东八区，领先 UTC 8 小时，所以北京时间=UTC+08:00。

6. 进入系统视图。

使用 system view 命令可以进入系统视图，在系统视图下可以配置设备名称、接口、协议等。

```
<Huawei>system-view //进入系统视图
Enter system view, return user view with Ctrl+Z.
#提示：如需返回用户视图，请按<Ctrl>+<Z>键
```

7. 修改设备名称。

配置设备时，为了便于区分，往往给设备定义不同的名称。例如，依照实验拓扑图，修改设备名称为 Switch。

```
[Huawei]sysname Switch //修改交换机的设备名称为Switch
#提示：在系统视图中执行此命令
[Switch]
```

8. 配置登录标语信息。

配置登录标语信息以用于提示或显示登录警告。执行 header shell information 命令以配置登录信息。

```
[Switch]header shell information "Unauthorized login prohibited."
//设置登录标语信息
#提示：在系统视图中执行此命令
```

退出命令行界面，再次登录命令行界面，查看登录信息是否已被修改。

```
[Switch]quit //退出系统视图
<Switch>quit //退出用户视图
  Configuration console exit, please press any key to log on
Unauthorized login prohibited. //显示登录信息
<Switch>
```

9. 配置 Console 接口参数。

默认情况下，通过 Console 接口登录时无须密码，任何人都可以直接连接设备进行配置。为避免由此带来的风险，可以将 Console 接口的登录方式配置为密码认证方式，并设置密码为"Huawei@123"。

空闲超时指的是经过没有任何操作的一定时间后，系统会自动退出该配置界面，再次登录时会根据系统要求，提示输入密码以进行验证。

设置空闲超时为 20 分钟，默认为 10 分钟。

```
[Switch]user-interface console 0//进入Console接口配置视图
[Switch-ui-console0]authentication-mode password//设置密码认证
[Switch-ui-console0]set authentication password cipher Huawei@123
[Switch-ui-console0] idle-timeout 20 //设置空闲超时为20分钟
```

退出系统，并使用新配置的密码登录系统。需要注意的是，在交换机第一次启动时，也需要配置密码。

```
[Switch-ui-console0]return//直接返回用户视图
<Switch>Save//保存配置命令
The current configuration (excluding the configurations of unregistered boards or cards)
will be written to flash:/vrpcfg.zip.
#提示：系统提示当前配置将被保存到flash:/vrpcfg.zip文件中，并会覆盖原有文件，要求确认
Are you sure to continue?[Y/N]y//输入yes以确认保存操作
```

```
     Now saving the current configuration to the slot 0.Apr  2 2000 00:35:41
SWB %%01CFM/4/SAVE(s)[6]:The user chose Y when deciding whether to save the configuration to
the device...
     Save the configuration successfully.
     #提示：系统提示保存成功
<Switch>quit  //退出用户视图
```

 任务总结与思考

本任务重点讲述交换机的基本配置命令，主要包括时间设置、设备命名、登录信息、Console 接口密码认证等，同时讲解交换机的多种视图模式及功能，需要熟悉掌握命令行帮助和自动补全功能。

思考以下两个问题。

1. 交换机常见的视图模式有哪些？在每个视图中可以完成什么功能？
2. 交换机 Console 接口登录认证方式有几种？如何进行配置？

 知识补给

2.2.1　常用的命令行视图

常用命令行视图名称及功能如表 2-2-1 所示。

表 2-2-1　常用命令行视图名称及功能

常用视图名称	进入视图	视图功能
用户视图	用户从终端成功登录至设备即进入用户视图。 　　　　`<HUAWEI>`	在用户视图中，用户可以完成查看运行状态和统计信息等功能

系统视图	在用户视图中，输入命令 `system-view` 后按 Enter 键，进入系统视图。
	　　　`<HUAWEI>system-view`
	　`Enter system view, return user view with Ctrl+Z.`
	　　　　`[HUAWEI]`
		在系统视图中，用户可以配置系统参数，以及通过该视图进入其他的功能配置视图

接口视图	使用 `interface` 命令并指定接口类型及接口编号进入相应的接口视图。
	　　`[HUAWEI] interface EthernetX/Y/Z`
	　　　　`[HUAWEI-EthernetX/Y/Z]`
	X/Y/Z 为需要配置的接口的编号，分别对应堆叠 ID/子卡号/接口序号。
	注：上述举例中的 Ethernet 接口仅为示意
		配置接口参数的视图称为接口视图。在该视图中可以设置接口的物理属性、链路层特性及 IP 地址等重要参数

命令行提示符 "HUAWEI" 是默认的主机名（sysname）。通过提示符可以判断当前所处的视图模式。例如，<>表示用户视图，[ ]表示除用户视图以外的其他视图。

用户可以在任意视图中输入！（或#）加字符串，此时的用户输入将全部（包括！和#在内）

作为系统的注释行内容,不会产生对应的配置信息。

2.2.2 退出命令行视图

执行 quit 命令,可从当前视图返回上一层视图。

例如,执行 quit 命令退出 AAA 视图并返回系统视图,再执行 quit 命令退出系统视图并返回用户视图。

```
[HUAWEI-aaa] quit//退出AAA视图
[HUAWEI] quit//退出系统视图
<HUAWEI>
```

如果需要从 AAA 视图直接返回用户视图,则可以在键盘上按组合键<Ctrl> + <Z>或执行 return 命令。

按组合键<Ctrl> + <Z>直接返回用户视图。

```
[HUAWEI-aaa]            //按<Ctrl> + <Z>直接返回用户视图
<HUAWEI>
```

执行 return 命令直接返回用户视图。

```
[HUAWEI-aaa] return//返回用户视图
<HUAWEI>
```

2.2.3 帮助功能和命令自动补全功能

在系统中输入命令时,? 是通配符,Tab 键是自动补全命令的快捷键。

```
<Huawei>display ?// 显示以display开头的命令
  aaa                        AAA
  access-author              Access user author
  access-context             Access user context
  access-user                User access
  accounting-scheme          Accounting scheme
  acl                        ACL status and configuration information
  alarm                      Alarm
  als                        Set automatic laser shutdown
  anti-attack                Specify anti-attack configurations
  arp                        Display ARP entries
  arp-limit                  Display the number of limitation
  as                         Access Switch
  assistant                  Assistant
  associate-user             Associate user
  authentication             Authentication
  authentication-profile     Authentication profile
  authentication-scheme      Authentication scheme
  authorization-scheme       Display AAA authorization scheme
  ---- More ----
```

在输入字符后输入?,可查看以输入字符开头的命令。例如,输入 dis?,系统将输出所有以 dis 开头的命令。在输入的字符后增加空格，然后再输入？，这时系统将尝试识别输入的信息所对应的命令，然后输出该命令的其他参数。例如，输入 dis？，如果只有 display 命令是以 dis 开头的，那么系统将输出 display 命令的参数；如果还有以 dis 开头的其他命令，则系统将报错。

另外，可以使用键盘上的 Tab 键补全命令，如输入 dis 后，按 Tab 键可以将命令补全为 display。如果有多个以 dis 开头的命令，则按 Tab 键后，可在多个命令之间循环切换。

在不发生歧义的情况下，命令可以使用简写，如 display 可以简写为 dis 或 disp 等，interface 可以简写为 int 或 inter 等。

2.3 隔离部门网络

隔离部门网络

➢ 任务情景

宇信公司的局域网已经搭建完成，为了提高网络的性能和服务质量，现在需要将财务部门和业务部门的网络隔离，使这两个部门使用不同的网段。小李作为网络管理员，为了保证两个部门相对独立，需要将属于同一部门的交换机端口划分到对应的 VLAN 中，从而使部门之间的数据互不干扰，且也使各自的通信效率不受影响。

➢ 任务分析

- ➢ 了解 VLAN 简介；
- ➢ 掌握 VLAN 划分；
- ➢ 学会验证 VLAN。

➢ 实施准备

1. HUAWEI S3700-26C-HI 交换机 1 台；
2. PC 4 台；
3. Console 通信线缆 1 根；
4. 以太网线缆 4 根。

➢ 实施步骤

1. 按照如图 2-3-1 所示网络拓扑结构，使用以太网线缆连接 PC 和交换机。
2. 交换机 VLAN、计算机 IP 地址及子网掩码划分如表 2-3-1 所示。

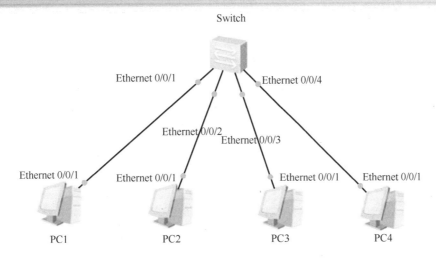

图 2-3-1　网络拓扑结构

表 2-3-1　交换机 VLAN、计算机 IP 地址及子网掩码划分

设备	端口成员	IP 地址	子网掩码	部门
Switch 的 VLAN 10	1～2			财务部
Switch 的 VLAN 20	3～4			业务部
PC1	Ethernet0/0/1	192.168.1.11	255.255.255.0	财务部
PC2	Ethernet0/0/2	192.168.1.12	255.255.255.0	财务部
PC3	Ethernet0/0/3	192.168.1.13	255.255.255.0	业务部
PC4	Ethernet0/0/4	192.168.1.14	255.255.255.0	业务部

小贴士

因为二层交换机不具备路由功能，所以二层交换机在同一 VLAN 中的端口能够通信，而在不同 VLAN 中的端口不能通信。因为三层交换机具备路由功能，所以可以在不同 VLAN 中的端口之间实现通信。

3．由于分配给每台计算机的 IP 都属于 192.168.1.0/24 网络，所以计算机之间能够通信。例如，划分 VLAN 前分别测试 PC1 到 PC2，以及 PC1 到 PC3 的连通性，如图 2-3-2 所示。

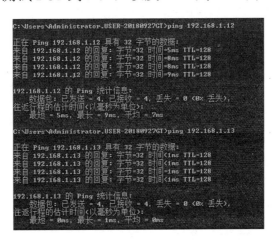

图 2-3-2　连通性测试

4. 在交换机上进行配置前，首先需要检查设备环境。如果使用的设备是空配置则可以直接进行配置；如果使用的设备包含原有配置，则需清除设备上的原有配置。操作步骤如下。

```
<HUAWEI>reset saved-configuration  // 删除闪存中的配置文件
This will delete the configuration in the flash memory.
The device configurations will be erased to reconfigure.
Are you sure? (y/n)[n]:y// 该设备配置即将被重置。是否确定（是/否）：是
 Clear the configuration in the device successfully.// 清除设备配置成功
<HUAWEI>reboot// 重启设备
Info: The system is now comparing the configuration, please wait.
Warning: All the configuration will be saved to the next startup configuration. Continue ?
[y/n]:n
// 警告：所有配置即将保存到下次启动配置。继续？[是/否]：否
System will reboot! Continue ? [y/n]:y// 系统即将重启！继续？[是/否]：是
Info: system is rebooting ,please wait...
```
#提示：清除配置是一项危险操作，需要在管理员监督下进行

5. 配置交换机主机名为 Switch。

```
<Huawei>system-view      //进入系统视图
Enter system view, return user view with Ctrl+Z.
```
#提示：如需返回用户视图，请按<Ctrl>+<Z>键
```
[Huawei] sysname Switch//把交换机的默认设备名称Huawei修改为Switch
[Switch]
```
#提示：修改生效，交换机设备名称已变为Switch

6. 在交换机 Switch 上创建 VLAN10、VLAN20，将 VLAN10 命名为 CAIWU，将 VLAN20 命名为 YEWU，并将 GE0/0/1～GE0/0/2 接口划分到 VLAN10，将 GE0/0/3～GE0/0/4 接口划分到 VLAN20。

```
[Switch]vlan 10        //创建VLAN10
[SWITCH-Vlan10]description CAIWU//命名VLAN10为CAIWU
```
#提示：为VLAN命名可以帮助管理员了解VLAN的属性及成员
```
[Switch-A-Vlan10]quit //退出VLAN配置视图
[Switch]vlan 20       //创建VLAN20
[Switch-Vlan20]description YEWU//命名VLAN20为YEWU
[Switch-Vlan20]quit//退出VLAN配置视图
[Switch]port-group pg1//创建端口组pg1
[SWitch-port-group-pg1] group-member Ethernet0/0/1 to Ethernet0/0/2
 //把接口Ethernet0/0/1～Ethernet0/0/2加入端口组
[Switch -port-group-pg1]port link-type access   //设置接口Ethernet0/0/1和Ethernet0/0/2的端口类型为access
[Switch -port-group-pg1] port default vlan 10
//设置接口Ethernet0/0/1和Ethernet0/0/2加入VLAN10
```
#提示：以下命令行为系统自动执行，无须人为输入，执行结果为依次转换GE0/0/1、GE0/0/2端口为access模式，并加入VLAN10
```
[Switch-Ethernet0/0/1]port default vlan 10
[Switch-Ethernet0/0/2]port default vlan 10
[Switch-port-group]quit//退出接口配置视图
```

```
[Switch]port-group pg2//创建端口组pg2
[SWitch-port-group-pg2] group-member Ethernet0/0/3 to Ethernet0/0/4
//把接口Ethernet0/0/3~Ethernet0/0/4加入端口组
[Switch-port-group-pg2]port link-type access
//设置接口Ethernet0/0/3、Ethernet0/0/4的端口类型为access
   [Switch-port-group]port default vlan 20   //将接口Ethernet0/0/3、Ethernet0/0/4划分到
VLAN20
   #提示：以下命令行为系统自动执行，无须人为输入，执行结果为依次转换Ethernet0/0/3~Ethernet0/0/4
端口为access模式，并加入VLAN20
   [Switch-Ethernet0/0/3]port default vlan 20
   [Switch-Ethernet0/0/4]port default vlan 20
   [Switch-port-group]quit//退出接口配置视图
```

7. 验证交换机 VLAN 配置。

```
[Switch]display vlan// 显示VLAN相关信息
The total number of vlans is : 3
--------------------------------------------------------------------
U: Up;          D: Down;         TG: Tagged;         UT: Untagged;
MP: Vlan-mapping;                ST: Vlan-stacking;
#: ProtocolTransparent-vlan;    *: Management-vlan;
--------------------------------------------------------------------
VID  Type    Ports
1    common  UT:
Eth0/0/5(D)      Eth0/0/6(D)      Eth0/0/7(D)     Eth0/0/8(D)     Eth0/0/9(U)
Eth0/0/10(D)     Eth0/0/11(D)     Eth0/0/12(D)    Eth0/0/13(D)    Eth0/0/14(D)
                 Eth0/0/15(D)     Eth0/0/16(D)    Eth0/0/17(D)    Eth0/0/18(D)
                 Eth0/0/19(D)     Eth0/0/20(D)    Eth0/0/21(D)    Eth0/0/22(D)
                 GE0/0/1(U)       GE0/0/2(D)
10   common  UT:Eth0/0/1(U)   Eth0/0/2(D)
             TG:GE0/0/1(U)
20   common  UT:Eth0/0/3(D)   Eth0/0/4(D)
             TG:GE0/0/1(U)
VID  Status  Property     MAC-LRN Statistics Description
--------------------------------------------------------------------
1    enable  default      enable   disable    VLAN 0001
10   enable  default      enable   disable    CAIWU
20   enable  default      enable   disable    YEWU         [Switch]
```

8. 配置完成后，测试网络连通性。

使用 ping 命令从 PC1 向 PC2 发送 4 个测试数据包，收到 PC2 的应答信息；使用 ping 命令从 PC1 向 PC3 发送 4 个测试数据包，收到 PC3 的应答信息。划分 VLAN 后分别测试 PC1 到 PC3，以及 PC1 到 PC3 的连通性，如图 2-3-3 所示。测试结果证明已经实现财务部门与业务部门的网络隔离。

图 2-3-3 连通性测试

任务总结与思考

本任务重点讲述如何在交换机上划分 VLAN 的方法，同一局域网内的 PC，只有划分到相同 VLAN 中才可以通信，不同 VLAN 中的 PC 不能通信，从而有效地实现网络隔离，保护通信数据。

思考以下两个问题。
1. 如何批量创建多个 VLAN？
2. 有几种划分 VLAN 的方式？

知识补给

2.3.1 VLAN 简介

VLAN（Virtual Local Area Network）即虚拟局域网，是将一个物理的 LAN 在逻辑上划分成多个广播域的通信技术。

以太网是一种基于 CSMA/CD（Carrier Sense Multiple Access/Collision Detection）的共享通信介质的数据网络通信技术。以太网中的主机数目较多会导致冲突严重、广播泛滥、性能显著下降，甚至产生网络不可用等问题。通过交换机实现 LAN 互连虽然可以解决冲突严重的问题，但仍然不能隔离广播报文和提升网络质量。

在这种情况下，VLAN 技术出现了，这种技术可以把一个局域网划分成多个逻辑的 VLAN，每个 VLAN 都是一个广播域，VLAN 内的主机之间的通信就和在一个 LAN 内一样，而 VLAN 之间则不能直接互通，这样广播报文就被限制在同一个 VLAN 内。

2.3.2 VLAN 划分

划分 VLAN 的方式有基于接口、基于 MAC 地址、基于子网、基于协议、基于匹配策略（MAC 地址、IP 地址、端口）。

➢ 基于接口划分 VLAN：根据交换机接口分配 VLAN。该方式配置简单，可以用于各种场景，常用于位置比较固定的网络。

➢ 基于 MAC 地址划分 VLAN：根据报文的源 MAC 地址分配 VLAN。该方式经常用于用户位置变化、不需要重新配置 VLAN 的场景，适用于位置经常移动但网卡不经常更换的小型网络，如移动 PC。

➢ 基于子网划分 VLAN：根据报文的源 IP 地址分配 VLAN。该方式一般用于对同一网段的用户进行统一管理的场景，适用于对安全需求不高，但对移动性和简易管理需求较高的场景。例如，一台 PC 配置多个 IP 以分别访问不同网段的服务器，以及 PC 切换 IP 后要求 VLAN 自动切换等场景。

➢ 基于协议划分 VLAN：根据数据帧所属的协议类型分配 VLAN。该方式经常用于对具有相同应用或服务的用户，进行统一管理的场景，适用于需要同时运行多协议的网络。

➢ 基于匹配策略划分 VLAN：根据指定的策略（如匹配报文的源 MAC、源 IP 和端口）分配 VLAN ID。该方式适用于对安全性要求比较高的场景。

2.3.3 VLAN 操作命令

1. 批量创建 VLAN。

在系统视图中执行 vlan batch 命令，以批量创建 VLAN。

➢ 批量创建 10 个连续的 VLAN：VLAN11～VLAN20。

```
<HUAWEI> system-view//进入系统视图
[HUAWEI] vlan batch 11 to 20 //批量创建VLAN11～VLAN20
```

➢ 批量创建 10 个不连续的 VLAN：VLAN10、VLAN15～VLAN19、VLAN25、VLAN28～VLAN30。

```
<HUAWEI>system-view//进入系统视图
[HUAWEI] vlan batch 10 15 to 19 25 28 to 30//批量创建VLAN10、VLAN15～VLAN19、VLAN25、VLAN28～VLAN30
```

小贴士

批量创建不连续的 VLAN 时，一次最多可以输入 10 个不连续的 VLAN 或 VLAN 段，超过 10 个后可以多次使用该命令进行配置。例如，vlan batch 10 15 to 19 25 28 to 30 一共是 4 个不连续的 VLAN 段。

2. 批量端口加入 VLAN。

```
<HUAWEI>system-view//进入系统视图
```

```
[HUAWEI] port-group pg1//创建端口组pg1
[HUAWEI-port-group-pg1] group-member Ethernet0/0/1 to Ethernet0/0/5
//把接口Ethernet0/0/1~Ethernet0/0/5加入端口组
[HUAWEI-port-group-pg1] port link-type access
//批量修改端口Ethernet0/0/1~Ethernet0/0/5的链路类型为access
[HUAWEI-port-group-pg1] port default vlan 10
//批量把端口Ethernet0/0/1~Ethernet0/0/5加入VLAN10
```

3. 删除 VLAN。

设备支持单个删除 VLAN 和批量删除 VLAN 两种方式。

> 单个删除 VLAN10。

```
<HUAWEI> system-view//进入系统视图
[HUAWEI] undo vlan 10//删除VLAN10
```

> 批量删除 VLAN10~VLAN20。

```
<HUAWEI> system-view//进入系统视图
[HUAWEI] undo vlan batch 10 to 20//删除VLAN10~VLAN20
```

2.4 相同 VLAN 中的通信

相同 VLAN 中的通信

> **任务情景**

宇信公司有财务部、市场部等部门，在不同楼层内都有财务部门和市场部门的员工计算机，为了方便公司管理，并使网络更加安全、便捷，领导想让管理员小李组建公司局域网络，以使各部门内的主机可以通信，但基于安全的考虑，需禁止各部门互相访问。

> **任务分析**

> 了解 VLAN 划分优点；
> 了解以太网接口类型；
> 掌握相同 VLAN 技术。

> **实施准备**

1. HUAWEIS3700-26C-HI 交换机 2 台；
2. PC 4 台；
3. Console 通信线缆 1 根；
3. 以太网线缆 5 根。

> **实施步骤**

1. 按照如图 2-4-1 所示网络拓扑结构，使用以太网线缆连接终端和设备。
2. 交换机 VLAN 划分如表 2-4-1 所示。

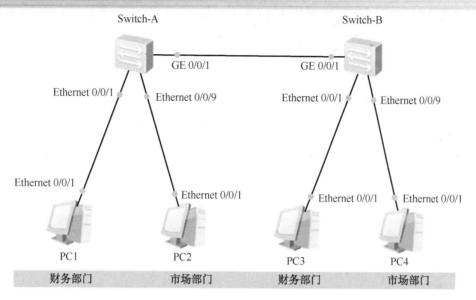

图 2-4-1　网络拓扑结构

表 2-4-1　交换机 VLAN 划分

设备	端口成员	IP 地址	子网掩码	备注
Switch-A 的 VLAN 10	1～5	无	无	无
Switch-A 的 VLAN 20	6～10	无	无	无
Switch-A	GE0/0/1（Trunk）	无	无	无
Switch-B 的 VLAN 10	1～5	无	无	无
Switch-B 的 VLAN 20	6～10	无	无	无
Switch-B	GE0/0/1（Trunk）	无	无	无
PC1	Switch-A 的 Ethernet0/0/1	192.168.1.10	255.255.255.0	财务部门
PC2	Switch-A 的 Ethernet0/0/9	192.168.1.20	255.255.255.0	市场部门
PC3	Switch-B 的 Ethernet0/0/1	192.168.1.30	255.255.255.0	财务部门
PC4	Switch-B 的 Ethernet0/0/9	192.168.1.40	255.255.255.0	市场部门

3．配置交换机 A 的主机名称为 Switch-A，创建 VLAN10 和 VLAN20，并将相应的端口分配到对应的 VLAN 中。

```
<Huawei>system-view //进入系统视图
Enter system view, return user view with Ctrl+Z.
#提示：如需返回用户视图，请按组合键<Ctrl>+<Z>
[Huawei]sysname Switch-A//把交换机的默认设备名称Huawei修改为Switch-A
[Switch-A]
#提示：修改生效，交换机设备名称已变为Switch-A
[Switch-A]vlan 10     //创建VLAN10
#提示：为VLAN命名可以帮助管理员了解VLAN的属性及成员
[Switch-A-Vlan10]description CAIWU     //命名VLAN10为CAIWU
[Switch-A-Vlan10]quit //退出VLAN配置视图
[Switch-A]vlan 20     //创建VLAN20
[Switch-A-Vlan20]description SHICHANG      //命名VLAN20为SHICHANG
[Switch-A-Vlan20]quit //退出VLAN配置视图
```

```
[Switch-A]port-group 10    //创建端口组10
[Switch-A-port-group-10] group-member Ethernet0/0/1 to Ethernet0/0/5
    //把接口Ethernet0/0/1~Ethernet0/0/5加入端口组10
[Switch-A-port-group-10] port link-type access
//批量修改端口Ethernet0/0/1~Ethernet0/0/5的链路类型为access
[Switch-A-port-group-10] port default vlan 10
//批量把端口Ethernet0/0/1~Ethernet0/0/5加入VLAN10
#提示：此命令行为系统自动执行，无须人为输入，以下5行情况相同
[Switch-A-Ethernet0/0/1]port default vlan 10
[Switch-A-Ethernet0/0/2]port default vlan 10
[Switch-A-Ethernet0/0/3]port default vlan 10
[Switch-A-Ethernet0/0/4]port default vlan 10
[Switch-A-Ethernet0/0/5]port default vlan 10
[Switch-A]port-group 20    //创建端口组20
[Switch-A-port-group-20] group-member Ethernet0/0/6 to Ethernet0/0/10
 //把接口Ethernet0/0/6~Ethernet0/0/10加入端口组20
[Switch-A-port-group-20] port link-type access
//批量修改端口Ethernet0/0/6~Ethernet0/0/10的链路类型为access
[Switch-A-port-group-20]port default vlan 20
//批量把端口Ethernet0/0/6~Ethernet0/0/10加入VLAN20
#提示：此命令行为系统自动执行，无须人为输入，以下5行情况相同
[Switch-A-Ethernet0/0/6]port default vlan 20
[Switch-A-Ethernet0/0/7]port default vlan 20
[Switch-A-Ethernet0/0/8]port default vlan 20
[Switch-A-Ethernet0/0/9]port default vlan 20
[Switch-A-Ethernet0/0/10]port default vlan 20
[Switch-A-port-group]quit //退出接口配置视图
```

4．配置交换机B的主机名称为Switch-B，创建VLAN10和VLAN20，并将相应的端口分配到对应VLAN中。

```
<Huawei>system-view //进入系统视图
Enter system view, return user view with Ctrl+Z.
#提示：如需返回用户视图，请按组合键<Ctrl>+<Z>
[Huawei] sysname Switch-B//将交换机的默认设备名称Huawei修改为Switch-B
[Switch-B]
#提示：修改生效，交换机设备名称已改为Switch-B
[Switch-B]vlan 10    //创建VLAN10
Info: This operation maytake a few seconds. Please wait for a moment...done.
#提示：系统提示此操作需要几秒钟，请等待完成后再继续操作
[Switch-B-Vlan10]description CAIWU   //命名VLAN10为CAIWU
#提示：为VLAN命名可以帮助管理员了解VLAN的属性及成员
[Switch-B-Vlan10]quit //退出VLAN配置视图
[Switch-B]vlan 20    //创建VLAN20
[Switch-B-Vlan20]description SHICHANG    //命名VLAN20为SHICHANG
[Switch-B-Vlan20]quit //退出VLAN配置视图
[Switch-B]port-group 10//创建端口组10
[Switch-B-port-group-10] group-member Ethernet0/0/1 to Gabitethernet0/0/5
 //把接口Ethernet0/0/1~Ethernet0/0/5加入端口组10
[Switch-B-port-group-10] port link-type access
```

```
//批量修改端口Ethernet0/0/1~Ethernet0/0/5的链路类型为access
[Switch-B-port-group-10] port default vlan 10
//批量把端口Ethernet0/0/1~Ethernet0/0/5加入VLAN10
#提示：此命令行为系统自动执行，无须人为输入，以下5行情况相同
[Switch-B-Ethernet0/0/1]port default vlan 10
[Switch-B-Ethernet0/0/2]port default vlan 10
[Switch-B-Ethernet0/0/3]port default vlan 10
[Switch-B-Ethernet0/0/4]port default vlan 10
[Switch-B-Ethernet0/0/5]port default vlan 10
[Switch-B]port-group 20    //创建端口组20
[Switch-B-port-group-20] group-member Ethernet0/0/6 to Ethernet0/0/10
//把接口Ethernet0/0/6~Ethernet0/0/10加入端口组20
[Switch-B-port-group-20] port link-type access
//批量修改端口Ethernet0/0/6~Ethernet0/0/10的链路类型为access
[Switch-B-port-group-20]port default vlan 20
//批量把端口Ethernet0/0/6~Ethernet0/0/10加入VLAN20
#提示：此命令行为系统自动执行，无须人为输入，以下5行情况相同
[Switch-B-Ethernet0/0/6]port default vlan 20
[Switch-B-Ethernet0/0/7]port default vlan 20
[Switch-B-Ethernet0/0/8]port default vlan 20
[Switch-B-Ethernet0/0/9]port default vlan 20
[Switch-B-Ethernet0/0/10]port default vlan 20
[Switch-B-port-group]quit //退出接口配置视图
```

5．验证交换机 A、交换机 B 的 VLAN 配置。

```
The total number of vlans is : 3
--------------------------------------------------------------------------
U: Up;          D: Down;        TG: Tagged;         UT: Untagged;
MP: Vlan-mapping;               ST: Vlan-stacking;
#: ProtocolTransparent-vlan;    *: Management-vlan;
--------------------------------------------------------------------------
VID  Type    Ports
1    common  UT:Eth0/0/11(D)   Eth0/0/12(D)   Eth0/0/13(D)   Eth0/0/14(D)
                Eth0/0/15(D)   Eth0/0/16(D)   Eth0/0/17(D)   Eth0/0/18(D)
                Eth0/0/19(D)   Eth0/0/20(D)   Eth0/0/21(D)   Eth0/0/22(D)
                GE0/0/1(U)     GE0/0/2(D)
10   common  UT:Eth0/0/1(U)    Eth0/0/2(D)    Eth0/0/3(D)    Eth0/0/4(D)
                Eth0/0/5(D)
             TG:GE0/0/1(U)
20   common  UT:Eth0/0/6(D)    Eth0/0/7(D)    Eth0/0/8(D)    Eth0/0/9(U)
                Eth0/0/10(D)
             TG:GE0/0/1(U)
VID  Status  Property    MAC-LRN  Statistics  Description
--------------------------------------------------------------------------
1    enable  default     enable   disable     VLAN 0001
10   enable  default     enable   disable     CAIWU
20   enable  default     enable   disable     SHICHANG
```

6．测试网络的连通性。

在 PC1 上分别 ping PC3 和 PC4 的 IP 192.168.1.30 和 192.168.1.40。分析连通性测试结果

（见图 2-4-2）可知，此时网络不连通。

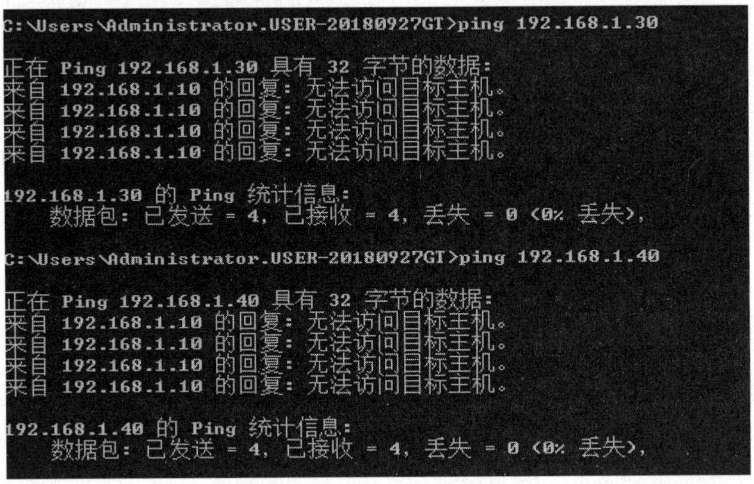

图 2-4-2　连通性测试结果

小贴士

跨交换机时，相同 VLAN 的计算机之间无法通信，主要是因为没有配置交换机级联的 Trunk 端口。

7．配置交换机 A 的 Trunk 端口。

```
[Swithc-A]interface GigabitEthernet0/0/1    //进入端口GE0/0/1
[Swithc-A-GigabitEthernet0/0/1]port link-type trunk
//配置端口为Trunk
 [Swithc-A-GigabitEthernet0/0/1]port trunk allow-pass vlan 10 20
//华为交换机默认不支持其他VLAN通过，但VLAN1除外，所以要允许Trunk端口通过VLAN10和VLAN20
Info: This operation may takea few seconds. Please wait for a moment...done.
#提示：系统提示此操作需要花费几秒钟，请耐心等待
[Swithc-A-Ethernet0/0/24]quit       //退出端口配置视图，返回系统视图
[Swithc-A]quit                      //退出系统视图
<Swithc-A>save                      //返回用户视图，并保存当前配置
```

8．配置交换机 B 的 Trunk 端口。

```
[Swithc-B]interface GigabitEthernet0/0/1    //进入端口GE0/0/1
[Swithc-B-GigabitEthernet0/0/1]port link-type trunk
//配置接口为Trunk
 [Swithc-B-GigabitEthernet0/0/1]port trunk allow-pass vlan 10 20
//华为交换机默认不支持其他VLAN通过，但VLAN1除外，所以要允许Trunk端口通过VLAN10和VLAN20
Info: This operation may takea few seconds. Please wait for a moment...done.
#提示：系统提示此操作需要花费几秒钟，请耐心等待
[Swithc-B-GigabitEthernet0/0/1]quit //退出端口配置视图，返回系统视图
[Swithc-B]quit      //退出系统视图
<Swithc-B>save      //返回用户视图，并保存当前配置
```

9．测试网络的连通性。

在 PC1 上 ping PC3 的 IP 192.168.1.30，此时网络连通，表明交换机间的 Trunk 链路已经

成功建立；ping PC4 的 IP 192.168.1.40，此时网络不通，表明相同 VLAN 间可以跨交换机设备进行通信，而不同 VLAN 间无法通信。连通性测试结果如图 2-4-3 所示。

图 2-4-3　连通性测试结果

任务总结与思考

本任务重点讲述同一 VLAN 中的用户跨接在不同的交换机上，通过配置交换机 Trunk 链路实现相同 VLAN 中的通信。

思考以下两个问题。

1．在 LAN 中使用 VLAN 技术的优点是什么？
2．如何恢复交换机接口上 VLAN 的默认配置？

知识补给

2.4.1　VLAN 主要的优点

1．防范广播风暴：将网络划分为多个 VLAN 可减少参与广播风暴的设备数量，每个 VLAN 都是一个独立的广播域，从而使每个广播域中的计算机数量大为减少，节省了带宽，提高了网络处理能力。

2．增强局域网的安全性：不同 VLAN 间的报文在传输时都是相互隔离的，即同一个 VLAN 内的用户不能和其他 VLAN 内的用户直接通信。可以将含有敏感数据的用户和网络中的其他用户进行隔离，从而减小泄露机密信息的可能性。

3．提高网络的健壮性：故障被限制在一个 VLAN 内，该 VLAN 内的故障不会影响其他 VLAN 的正常工作。

4．灵活构建虚拟工作组：用 VLAN 可以将不同的用户划分到不同的工作组，同一工作组的用户不必局限于某一固定的物理范围，网络构建和维护更方便灵活。

2.4.2 以太网接口类型

根据接口连接对象及对收发数据帧处理的不同，以太网接口可以划分为以下三种。

1．Access 接口。

Access 接口一般用于和不能识别 Tag 的用户终端（如用户主机、服务器等）相连，或者在不需要区分不同 VLAN 成员时使用。它只能收/发 Untagged 帧，且只能为 Untagged 帧添加唯一 VLAN 的 Tag。

2．Trunk 接口。

Trunk 接口一般用于连接交换机、路由器、AP 及可同时收/发 Tagged 帧和 Untagged 帧的语音终端。它可以允许多个 VLAN 的帧带 Tag 通过，但只允许一个 VLAN 的帧从该类接口上发出时不带 Tag 通过（剥除 Tag）。

3．Hybrid 接口。

Hybrid 接口既可以用于连接不能识别 Tag 的用户终端（如用户主机、服务器等）和网络设备（如 Hub、傻瓜交换机），也可以用于连接交换机、路由器，且可同时收/发 Tagged 帧和 Untagged 帧的语音终端、AP。它可以允许多个 VLAN 的帧带 Tag 通过，且允许从该类接口发出的帧根据需要配置某些 VLAN 的帧带 Tag 通过（不剥除 Tag）、某些 VLAN 的帧不带 Tag 通过（剥除 Tag）。

2.4.3 恢复接口上 VLAN 的默认配置

接口上 VLAN 的配置包括默认 VLAN 和接口加入的 VLAN（接口允许通过的 VLAN）两部分。在默认情况下，接口上 VLAN 的默认配置如下。

- 链路类型是 Access：默认 VLAN 为 VLAN1，接口以 Untagged 方式加入 VLAN1。
- 链路类型是 Trunk：默认 VLAN 为 VLAN1，接口以 Tagged 方式加入 VLAN1。
- 链路类型是 Hybrid：默认 VLAN 为 VLAN1，接口以 Untagged 方式加入 VLAN1。
- 链路类型是 Dot1q-tunnel：默认 VLAN 为 VLAN1，接口加入 VLAN1。

在接口视图下，首先执行 display this include-default | include link-type 命令查看当前接口的链路类型，然后使用以下方法恢复接口上 VLAN 的默认配置。

1．恢复 Access 或 Dot1q-tunnel 链路接口上 VLAN 的默认配置。

```
<HUAWEI> system-view         //进入系统视图
[HUAWEI] interface Ethernet 0/0/1 //进入Ethernet0/0/1接口视图
[HUAWEI-Ethernet0/0/1] undo port default vlan//撤销接口默认的VLAN
```

2．恢复 Trunk 链路接口上 VLAN 的默认配置。

```
<HUAWEI> system-view//进入系统视图
[HUAWEI] interface Ethernet 0/0/1//进入Ethernet0/0/1接口视图
[HUAWEI-Ethernet0/0/1] undo port trunk pvid vlan
```

```
                // 撤销Trunk接口设置的本征VLAN
[HUAWEI-Ethernet0/0/1] undo port trunk allow-pass vlan all
                // 撤销接口允许通过的VLAN流量
[HUAWEI-Ethernet0/0/1] port trunk allow-pass vlan 1
                // 设置接口允许通过VLAN1流量
```

3. 恢复 Hybrid 接口上 VLAN 的默认配置。

```
<HUAWEI> system-view      //进入系统视图
[HUAWEI] interface Ethernet 0/0/1  //进入Ethernet0/0/1接口视图
[HUAWEI-Ethernet0/0/1] undo port hybrid pvid vlan
 // 撤销Hybird接口设置的本征VLAN
[HUAWEI-Ethernet0/0/1] undo port hybrid vlan all
 //  撤销Hybrid接口允许通过的所有VLAN流量
[HUAWEI-Ethernet0/0/1] port hybrid untagged vlan 1
// 设置接口允许通过未达标的VLAN1流量
```

2.5 避免形成网络环路（STP 技术）

避免形成网络环路

➢ 任务情景

由于业务迅速发展和对网络可靠性要求的提高，宇信公司增加了设备之间的备份链路，希望在保证网络不间断的情况下能够达到最佳的工作效率，但是这样做难免会形成网络环路。若网络中存在网络环路，就可能引起广播风暴和 MAC 表项被破坏，为此管理员小李需要在网络中部署 STP 生成树协议，以预防形成网络环路。

➢ 任务分析

- ➢ 了解 STP 定义和目的；
- ➢ 了解 STP 收敛的过程；
- ➢ 掌握 STP 的配置方法。

➢ 实施准备

1．HUAWEIS3700-26C-HI 交换机 2 台；
2．PC 4 台；
3．Console 通信线缆 1 根；
4．以太网线缆 6 根。

➢ 实施步骤

1．按照如图 2-5-1 所示网络拓扑结构，使用线缆连接终端和设备。

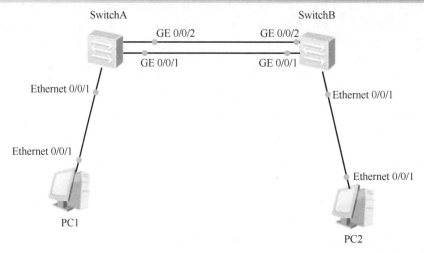

图 2-5-1 网络拓扑结构

2. 配置 PC 的 IP 地址及子网掩码，其分配表如表 2-5-1 所示。

表 2-5-1 PC 的 IP 地址及子网掩码分配表

PC 编号	所属 VLAN	IP 地址	子网掩码
PC1	10	192.168.10.1	255.255.255.0
PC2	20	192.168.20.1	255.255.255.0
PC3	10	192.168.10.2	255.255.255.0
PC4	20	192.168.20.2	255.255.255.0

3. 配置交换机 A 的主机名称为 SwitchA，创建 VLAN10 和 VLAN20，将相应的接口分配到对应的 VLAN 中，并配置交换机之间的互连端口为中继状态。

```
<Huawei>system-view      //进入系统视图
Enter system view, return user view with Ctrl+Z.
#提示：如需返回用户视图，请组合键按<Ctrl>+<Z>
[Huawei]sysname SwitchA//将交换机的默认设备名称Huawei修改为SwitchA
[SwitchA]
#提示：修改生效，交换机设备名称已变为SwitchA
[SwitchA]vlan batch 10 20//批量创建VLAN10、VLAN20
[SwitchA]port-group 10//创建端口组10
[SwitchA-port-group-10] group-member Ethernet0/0/1 to Ethernet0/0/5
 //把接口Ethernet0/0/1～Ethernet0/0/5加入端口组10
[SwitchA-port-group-10] port link-type access
//批量修改端口Ethernet0/0/1 ～ Ethernet0/0/5的链路类型为access
[SwitchA-port-group-10] port default vlan 10
//批量把端口Ethernet0/0/1 ～ Ethernet0/0/5加入VLAN10
#提示：此命令行由系统自动执行，无须人为输入，以下5行情况相同
[SwitchA-Ethernet0/0/1]port default vlan 10
[SwitchA-Ethernet0/0/2]port default vlan 10
[SwitchA-Ethernet0/0/3]port default vlan 10
[SwitchA-Ethernet0/0/4]port default vlan 10
[SwitchA-Ethernet0/0/5]port default vlan 10
[SwitchA]port-group 20//创建端口组20
[SwitchA-port-group-20] group-member Ethernet0/0/6 to Ethernet0/0/10
```

```
//把接口Ethernet0/0/6～Ethernet0/0/10加入端口组20
[SwitchA-port-group-20] port link-type access
//批量修改接口Ethernet0/0/6～Ethernet0/0/10的链路类型为access
[SwitchA-port-group-20]port default vlan 20
//批量把接口Ethernet0/0/6～Ethernet0/0/10加入VLAN20
#提示：此命令行由系统自动执行，无须人为输入，以下5行情况相同
[SwitchA-Ethernet0/0/6]port default vlan 20
[SwitchA-Ethernet0/0/7]port default vlan 20
[SwitchA-Ethernet0/0/8]port default vlan 20
[SwitchA-Ethernet0/0/9]port default vlan 20
[SwitchA-Ethernet0/0/10]port default vlan 20
[SwitchA-port-group]quit //退出接口配置视图
[SWitchA]int GigabitEthernet 0/0/1//进入接口GE0/0/1
[SWitchA-GigabitEthernet 0/0/1]port link-type trunk
//配置链路类型为Trunk（中继状态）
[SWitchA-GigabitEthernet 0/0/1]port trunk allow-pass vlan 10 20
//配置端口允许通过VLAN10、VLAN20的流量
[SWitchA]int GigabitEthernet 0/0/2    //进入接口GE0/0/2
[SWitchA-GigabitEthernet 0/0/2]port link-type trunk
//配置链路类型为Trunk（中继状态）
[SWitchA-GigabitEthernet 0/0/2]port trunk allow-pass vlan 10 20
//配置端口允许通过VLAN10、VLAN20的流量
```

4. 配置交换机 B 主机名称为 SwitchB，创建 VLAN10 和 VLAN20，将相应的接口分配到对应 VLAN 里，并配置交换机间的互连接口为中继状态。

```
Vlan<Huawei>system-view //进入系统视图
Enter system view, return user view with Ctrl+Z.
#提示：如需返回用户视图，请按组合键<Ctrl>+<Z>
[Huawei] sysname SwitchB//将交换机的默认设备名称Huawei修改为SwitchB
[SwitchB]
#提示：修改生效，交换机设备名称已变为SwitchB
[SwitchB]vlan  batch 10 20//批量创建VLAN10、VLAN20
[SwitchB]port-group 10//创建端口组10
[SwitchB-port-group-10] group-member Ethernet0/0/1 to Ethernet0/0/5
 //把接口Ethernet0/0/1～Ethernet0/0/5加入接口组10
[SwitchB-port-group-10] port link-type access
//批量修改接口Ethernet0/0/1～Ethernet0/0/5的链路类型为access
[SwitchB-port-group-10] port default vlan 10
//批量把接口Ethernet0/0/1 ～ Ethernet0/0/5加入VLAN10
#提示：此命令行系统自动执行，无须人为输入，以下5行情况相同
[SwitchB-Ethernet0/0/1]port default vlan 10
[SwitchB-Ethernet0/0/2]port default vlan 10
[SwitchB-Ethernet0/0/3]port default vlan 10
[SwitchB-Ethernet0/0/4]port default vlan 10
[SwitchB-Ethernet0/0/5]port default vlan 10
[SwitchB]port-group 20//创建端口组20
[SwitchB-port-group-20] group-member Ethernet0/0/6 to Ethernet0/0/10
 //把接口Ethernet0/0/6～Ethernet0/0/10加入端口组20
[SwitchB-port-group-20] port link-type access
//批量修改端口Ethernet0/0/6～Ethernet0/0/10的链路类型为access
[SwitchB-port-group-20]port default vlan 20
//批量把端口Ethernet0/0/6～Ethernet0/0/10加入VLAN20
```

```
                #提示：此命令行系统自动执行，无须人为输入，以下5行情况相同
                [SwitchB-Ethernet0/0/6]port default vlan 20
                [SwitchB-Ethernet0/0/7]port default vlan 20
                [SwitchB-Ethernet0/0/8]port default vlan 20
                [SwitchB-Ethernet0/0/9]port default vlan 20
                [SwitchB-Ethernet0/0/10]port default vlan 20
                [SwitchB-port-group]quit //退出接口配置视图
                [SWitchB]int GigabitEthernet 0/0/1    //进入接口GE0/0/1
                [SWitchB-GigabitEthernet 0/0/1]port link-type trunk
                //配置链路类型为Trunk（中继状态）
                [SWitchB-GigabitEthernet 0/0/1]port trunk allow-pass Vlan 10 20
                //配置端口允许通过VLAN10、VLAN20的流量
                [SWitchB]int GigabitEthernet 0/0/2    //进入接口GE0/0/2
                [SWitchB-GigabitEthernet 0/0/2]port link-type trunk
                //配置链路类型为Trunk（中继状态）
                [SWitchB-GigabitEthernet 0/0/2]port trunk allow-pass Vlan 10 20
                //配置端口允许通过VLAN10、VLAN20的流量
```

5. 验证交换机 A、交换机 B 的 VLAN 配置。

```
                [SwitchA]display vlan
                The total number of vlans is : 3
                U: Up;         D: Down;            TG: Tagged;           UT: Untagged;
                MP: Vlan-mapping;               ST: Vlan-stacking;
                #: ProtocolTransparent-vlan;    *: Management-vlan;
                --------------------------------------------------------------------
                VID  Type   Ports
                --------------------------------------------------------------------
                1    common  UT:Eth0/0/11(D)   Eth0/0/12(D)   Eth0/0/13(D)   Eth0/0/14(D)
                                Eth0/0/15(D)   Eth0/0/16(D)   Eth0/0/17(D)   Eth0/0/18(D)
                                Eth0/0/19(D)   Eth0/0/20(D)   Eth0/0/21(D)   Eth0/0/22(D)
                                GE0/0/1(U)     GE0/0/2(U)
                10   common  UT:Eth0/0/1(U)    Eth0/0/2(D)    Eth0/0/3(D)    Eth0/0/4(D)
                                Eth0/0/5(D)
                             TG:GE0/0/1(U)     GE0/0/2(U)
                20   common  UT:Eth0/0/6(D)    Eth0/0/7(D)    Eth0/0/8(D)    Eth0/0/9(U)
                                Eth0/0/10(D)
                             TG:GE0/0/1(U)     GE0/0/2(U)
                VID  Status  Property     MAC-LRN Statistics Description
                --------------------------------------------------------------------
                1    enable  default      enable   disable    VLAN 0001
                10   enable  default      enable   disable    VLAN 0010
                20   enable  default      enable   disable    VLAN 0020
```

6. 在交换机 SwitchA、SwitchB 上配置 STP 生成树协议。

```
                [SwitchA]stp mode stp   // 配置生成树协议模式为STP
                Info: This operation may take a few seconds. Please wait for a moment...done.
                #提示：系统提示此操作需要几秒时间，请等待运行完成后再继续操作
                [SWitchB]stp mode stp   // 配置生成树协议模式为STP
                Info: This operation may take a few seconds. Please wait for a moment...done.
                #提示：系统提示此操作需要几秒时间，请等待运行完成后再继续操作
```

7. 验证并查看 STP 信息。

```
                [SwitchA]display stp brief  //查看交换机SwitchA的STP配置概要信息
```

```
 MSTID    Port                          Role   STP State    Protection
   0      Ethernet0/0/1                 DESI   FORWARDING   NONE
   0      Ethernet0/0/9                 DESI   FORWARDING   NONE
   0      GigabitEthernet0/0/1          DESI   FORWARDING   NONE
   0      GigabitEthernet0/0/2          DESI   FORWARDING   NONE
```

```
[SWitchB]display stp brief     //查看交换机SwitchB的STP配置概要信息
 MSTID    Port                          Role   STP State    Protection
   0      Ethernet0/0/1                 DESI   FORWARDING   NONE
   0      Ethernet0/0/9                 DESI   FORWARDING   NONE
   0      GigabitEthernet0/0/1          ROOT   FORWARDING   NONE
   0      GigabitEthernet0/0/2          ALTE   DISCARDING   NONE
```

```
[SwitchA]display stp interface GigabitEthernet 0/0/1
 //查看GE0/0/1端口STP信息
 -------[CIST Global Info][Mode STP]-------
 CIST Bridge            :0.7858-6043-7090
 Config Times           :Hello 2s MaxAge 20s FwDly 15s MaxHop 20
 Active Times           :Hello 2s MaxAge 20s FwDly 15s MaxHop 20
 CIST Root/ERPC         :0.7858-6043-7090 / 0 (This bridge is the root)
 CIST RegRoot/IRPC      :0.7858-6043-7090 / 0
 CIST RootPortId        :0.0
 BPDU-Protection        :Disabled
 CIST Root Type         :Primary root       // 根类型：主根
 TC or TCN received     :32
 TC count per hello     :0
 STP Converge Mode      :Normal
 Share region-configuration :Enabled
 Time since last TC     :0 days 0h:14m:34s
 Number of TC           :25
 Last TC occurred       :Ethernet0/0/23
 ----[Port32(Ethernet0/0/23)][FORWARDING]----
 Port Protocol          :Enabled
 Port Role              :DesignatedPort     //端口角色：指定端口
 Port Priority          :128                //端口优先级：128
 Port Cost(Dot1T )      :Config=auto / Active=20000
 Designated Bridge/Port :0.7858-6043-7090 / 128.32
 Port Edged             :Config=default / Active=disabled
 Point-to-point         :Config=auto / Active=true
 Transit Limit          :6 packets/s
 Protection Type        :None
 Port STP Mode          :STP       //协议类型：STP
 Port Protocol Type     :Config=auto / Active=dot1s
 BPDU Encapsulation     :Config=stp / Active=stp
 PortTimes              :Hello 2s MaxAge 20s FwDly 15s RemHop 20
 TC or TCN send         :53
 TC or TCN received     :2
```

```
    BPDU Sent              :566
           TCN: 0, Config: 566, RST: 0, MST: 0
    BPDU Received          :5
           TCN: 2, Config: 3, RST: 0, MST: 0
  Last forwarding time: 2019/01/08 16:45:31 UTC+08:00
[SwitchA]display stp interface GigabitEthernet 0/0/2
//查看GE0/0/2端口STP信息
-------[CIST Global Info][Mode STP]-------
CIST Bridge            :0.7858-6043-7090
Config Times           :Hello 2s MaxAge 20s FwDly 15s MaxHop 20
Active Times           :Hello 2s MaxAge 20s FwDly 15s MaxHop 20
CIST Root/ERPC         :0.7858-6043-7090 / 0 (This bridge is the root)
CIST RegRoot/IRPC      :0.7858-6043-7090 / 0
CIST RootPortId        :0.0
BPDU-Protection        :Disabled
CIST Root Type         :Primary root      //根类型：主根
TC or TCN received     :32
TC count per hello     :0
STP Converge Mode      :Normal
Share region-configuration :Enabled
Time since last TC     :0 days 0h:14m:41s
Number of TC           :25
Last TC occurred       :Ethernet0/0/23
----[Port33(Ethernet0/0/24)][FORWARDING]----
 Port Protocol         :Enabled
 Port Role             :Designated Port     //端口角色：指定端口
 Port Priority         :128                 //端口优先级：128
 Port Cost(Dot1T )     :Config=auto / Active=20000
 Designated Bridge/Port :0.7858-6043-7090 / 128.33
 Port Edged            :Config=default / Active=disabled
 Point-to-point        :Config=auto / Active=true
 Transit Limit         :6 packets/s
 Protection Type       :None
 Port STP Mode         :STP          //协议类型：STP
 Port Protocol Type    :Config=auto / Active=dot1s
 BPDU Encapsulation    :Config=stp / Active=stp
 PortTimes             :Hello 2s MaxAge 20s FwDly 15s RemHop 20
 TC or TCN send        :52
 TC or TCN received    :0
 BPDU Sent             :567
           TCN: 0, Config: 567, RST: 0, MST: 0
 BPDU Received         :5
           TCN: 0, Config: 5, RST: 0, MST: 0
 Last forwarding time: 2019/01/08 16:45:31 UTC+08:00
[SwitchA] quit   //退出系统视图
<SwitchA> save   //保存当前配置
```

```
[SWitchB]display stp interface Ethernet 0/0/1
 //查看E0/0/1端口STP信息
  -------[CIST Global Info][Mode STP]-------
  CIST Bridge              :4096.7858-6043-70b0
  Config Times             :Hello 2s MaxAge 20s FwDly 15s MaxHop 20
  Active Times             :Hello 2s MaxAge 20s FwDly 15s MaxHop 20
  CIST Root/ERPC           :0.7858-6043-7090 / 20000
  CIST RegRoot/IRPC        :4096.7858-6043-70b0 / 0
  CIST RootPortId          :128.23 (Ethernet0/0/23)
  BPDU-Protection          :Disabled
  CIST Root Type           :Secondary root    //根类型：次根
  TC or TCN received       :112
  TC count per hello       :0
  STP Converge Mode        :Normal
  Share region-configuration :Enabled
  Time since last TC       :0 days 0h:4m:51s
  Number of TC             :8
  Last TC occurred         :Ethernet0/0/23
  ----[Port23(Ethernet0/0/23)][FORWARDING]----
   Port Protocol           :Enabled
  Port Role                :Root Port    //端口角色：根端口
   Port Priority           :128
   Port Cost(Dot1T )       :Config=auto / Active=20000
   Designated Bridge/Port  :0.7858-6043-7090 / 128.32
   Port Edged              :Config=default / Active=disabled
   Point-to-point          :Config=auto / Active=true
   Transit Limit           :6 packets/s
   Protection Type         :None
   Port STP Mode           :STP          //协议类型：STP
   Port Protocol Type      :Config=auto / Active=dot1s
   BPDU Encapsulation      :Config=stp / Active=stp
   PortTimes               :Hello 2s MaxAge 20s FwDly 15s RemHop 0
   TC or TCN send          :1
   TC or TCN received      :29
   BPDU Sent               :3
         TCN: 1, Config: 2, RST: 0, MST: 0
   BPDU Received           :162
         TCN: 0, Config: 162, RST: 0, MST: 0
  Last forwarding time: 2000/04/02 00:26:43 UTC
---------------------------------------------------------------
[SWitchB]display stp interface Ethernet 0/0/1
 //查看接口E0/0/1的端口状态
  -------[CIST Global Info][Mode STP]-------
  CIST Bridge              :4096.7858-6043-70b0
  Config Times             :Hello 2s MaxAge 20s FwDly 15s MaxHop 20
  Active Times             :Hello 2s MaxAge 20s FwDly 15s MaxHop 20
  CIST Root/ERPC           :0.7858-6043-7090 / 20000
```

```
      CIST RegRoot/IRPC     :4096.7858-6043-70b0 / 0
      CIST RootPortId       :128.23 (Ethernet0/0/23)
      BPDU-Protection       :Disabled
      CIST Root Type        :Secondary root  //根类型：次根
      TC or TCN received    :112
      TC count per hello    :0
      STP Converge Mode     :Normal
      Share region-configuration :Enabled
      Time since last TC    :0 days 0h:6m:29s
      Number of TC          :8
      Last TC occurred      :Ethernet0/0/23
       ----[Port24(Ethernet0/0/24)][DISCARDING]----
       Port Protocol          :Enabled
       Port Role              :Alternate Port  //端口角色：替换端口
       Port Priority          :128             //端口优先级：128
       Port Cost(Dot1T )      :Config=auto / Active=20000
       Designated Bridge/Port :0.7858-6043-7090 / 128.33
       Port Edged             :Config=default / Active=disabled
       Point-to-point         :Config=auto / Active=true
       Transit Limit          :6 packets/s
       Protection Type        :None
       Port STP Mode          :STP    //协议类型：STP
       Port Protocol Type     :Config=auto / Active=dot1s
       BPDU Encapsulation     :Config=stp / Active=stp
       PortTimes              :Hello 2s MaxAge 20s FwDly 15s RemHop 0
       TC or TCN send         :0
       TC or TCN received     :28
       BPDU Sent              :3
            TCN: 0, Config: 3, RST: 0, MST: 0
       BPDU Received          :210
            TCN: 0, Config: 210, RST: 0, MST: 0
      #提示：命令行中加粗部分的信息表示当前端口的根类型、端口角色、端口优先级和STP协议类型
      [SwitchB]quit       //退出系统视图
      <SwitchB> save      //保存当前配置
```

任务总结与思考

本任务主要介绍当互联网中存在多条冗余备份链路时，通过在交换机上配置STP生成树协议，以解决互联网中可能存在的环路问题。

思考以下两个问题。

1. 在以太网交换网络中，由于冗余链路增加，可能对网络造成的危害有哪些？
2. 在交换机上配置了STP生成树协议后，交换机如何选取根桥、根接口和指定接口？

知识补给

2.5.1 STP 的定义和目的

1. 定义。

在以太网交换网络中，为了进行链路备份、提高网络可靠性，通常会使用冗余链路。但是使用冗余链路可能会在交换网络上产生环路，引发广播风暴及 MAC 地址表不稳定等故障，从而导致用户通信质量较差，甚至出现通信中断的情况。为解决以太网交换网络中的环路问题，人们开发了生成树协议（Spanning Tree Protocol，STP）。

运行 STP 的设备通过彼此交互信息发现网络中的环路，并有选择地对某个接口进行阻塞，最终将环形网络结构修剪成无环路的树形网络结构，以防止报文在环形网络中循环，并避免设备由于重复接收相同的报文造成处理能力下降。

2. 目的。

在以太网交换网中部署生成树协议后，如果网络中出现环路，则可利用生成树协议，通过拓扑计算实现以下功能。

- 消除环路：通过阻塞冗余链路，消除网络中可能存在的网络环路。
- 链路备份：在当前活动的路径发生故障时，激活冗余备份链路，恢复网络连通性。

2.5.2 STP 的收敛过程

收敛是指网络在初始化的一段时间内选举出作为根桥的交换机，在选举过程中，交换机端口可能需要进行不同的端口状态的转换，从而使所有交换机端口稳定地处于生成树端口角色状态，并使所有潜在的环路都被消除。收敛的过程需要耗费一定时间，这是因为在生成树形成过程中，各种状态转换都需要一定的时间来协调整个过程。

STP 的收敛过程有以下三个步骤。

步骤 1，选举根桥（根交换机）。

根桥是所有生成树路径开销计算的基础，用于防止形成环路的各种端口角色也是基于根桥而分配的。选举根桥在交换机完成启动时或在网络中检测到路径故障时触发。选举根桥是根据交换机的网桥 ID 进行的，其中网桥 ID=网桥优先级+VLAN ID，网桥 ID 最小者当选根桥。

步骤 2，选举根端口。

所有非根桥都产生一个根端口，即本网桥选定一个距根桥开销最小的端口作为根端口。

步骤 3，选举指定端口和非指定端口。

当交换机确定根端口后，还必须将剩余端口配置为指定端口（DP）或非指定端口（非 DP），以最终形成一个逻辑上无环路的网络拓扑。

交换网络中的每个网段都只能有一个指定端口。当两个非根端口的交换机端口连接到同一个 LAN 网段时，会发生竞争端口角色的情况。这两台交换机会交换生成树消息，以确定交换机的端口中哪个是指定端口，哪个是非指定端口。

一般而言，交换机端口是否配置为指定端口由网桥 ID 决定。不过，首要条件是具有到根桥的最低路径开销。只有当端口开销相等时，才考虑选择发送方的网桥 ID。

现代的交换机都是以往网桥产品的升级产品，所以本书中使用根桥代表根交换机。

2.5.3 常见 STP 操作

1. 开启 STP。

➢ 开启全局 STP。

在系统视图中执行命令 stp enable。

```
<HUAWEI>system-view
[HUAWEI] stp enable
```

➢ 开启接口 STP。

在接口视图中执行命令 stp enable。

```
<HUAWEI>system-view
[HUAWEI] interface Ethernet0/0/1
[HUAWEI-Ethernet0/0/1] stp enable
```

2. 关闭 STP。

➢ 关闭全局 STP。

在系统视图中执行命令 undo stp enable。

```
<HUAWEI> system-view
[HUAWEI] undo stp enable
```

➢ 关闭接口 STP。

在接口视图中执行命令 undo stp enable。

```
<HUAWEI>system-view
[HUAWEI] interface Ethernet0/0/1
[HUAWEI-Ethernet0/0/1] undo stp enable
```

3. 配置根桥和备份根桥。

交换机可以通过计算来自动确定生成树的根桥，用户也可以手动配置交换机指定生成树的根桥或备份根桥。

```
# 配置根桥
<HUAWEI>system-view
[HUAWEI] stp root primary
# 配置备份根桥
<HUAWEI>system-view
[HUAWEI] stp root secondary
```

4. 查看 STP 状态。

```
# 查看生成树的状态和统计信息摘要
<HUAWEI>display stp brief
```

```
MSTID     Port                    Role  STP State   Protection
  0       Ethernet0/0/21     DESI      FORWARDING      NONE
  0       Ethernet0/0/22     DESI      FORWARDING      NONE
  0       Ethernet0/0/23     DESI      FORWARDING      NONE
  0       Ethernet1/0/24     DESI      FORWARDING      NONE
```

5．查看根桥信息。

```
# 查看根桥的生成树状态信息
<HUAWEI>display stp bridge root
MSTID        Root ID            Root Cost Hello Max Forward Root Port
                                          Time Age  Delay
-----  -----------------------  --------- ----- --- ------- ---------
 0     61440.781d-ba56-f06c        0        2   20    15
```

思考与实训 2

一、选择题

1．交换机如何确定将帧转发到哪个端口？（　　）

  A．利用 MAC 地址表。 B．利用 ARP 地址表。

  C．读取源 ARP 地址。 D．读取源 MAC 地址。

2．一个 VLAN 可以看作一个（　　）。

  A．冲突域 B．广播域

  C．管理域 D．阻塞域

3．在 STP 中，当网桥的优先级相同时，以下（　　）将被选为根桥。

  A．拥有最小 MAC 地址的网桥 B．拥有最大 MAC 地址的网桥

  C．端口优先级数值最高的网桥 D．端口优先级数值最低的网桥

4．VLAN 的划分方法有（　　）。（多选题）

  A．基于协议 B．基于子网

  C．基于 MAC 地址 D．基于物理位置

二、填空题

1．本任务中的 HUAWEI S3700-26C-HI 对应 OSI 参考模型的_____层设备。

2．交换机的主要技术指标有_____、_____、_____、_____、_____和_____。

三、实训操作

假设搭建一个简单的公司网络场景，SW1 和 SW2 为楼层交换机，PC1 和 PC3 属于公司的部门 A，PC2 和 PC4 属于公司的部门 B，PC5 属于部门 A 和部门 B 的上级部门 C。在网络

规划中,部门 A 属于 VLAN 10,部门 B 属于 VLAN 20,部门 C 属于 VLAN 30。公司希望通过 VLAN 的划分和配置,使部门 A 和部门 B 均能够与部门 C 进行通信,但是要求部门 A 和部门 B 之间不能通信。PC 的 IP 地址表如表 2-1 所示,网络拓扑图如图 2-1 所示。

表 2-1 PC 的 IP 地址表

设备	IP 地址	子网掩码	网关
PC1	10.0.1.1	255.255.255.0	N/A
PC2	10.0.2.2	255.255.255.0	N/A
PC3	10.0.1.3	255.255.255.0	N/A
PC4	10.0.2.4	255.255.255.0	N/A
PC5	10.0.3.5	255.255.255.0	N/A

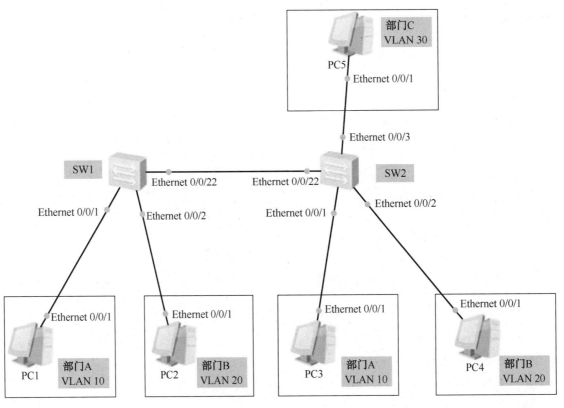

图 2-1 网络拓扑图

项目 3

三层业务互访

☆ 项目背景

宇信公司搬入新的办公园区后，小李在高工程师的指导下，根据各部门业务需要，规划并组建了各部门的业务网络，初步实现了绿色无纸办公。随着业务的深入，原来相互隔离的部门提出了互访的需求，小李在高工程师的指导下，购置了一批三层网络设备，包括 HUAWEIS5720 交换机和 HUAWEIAR2220E 路由器，用来满足不同业务网络互访的需求。

3.1 利用三层交换机实现部门网络互通

利用三层交换机
实现部门网络互通

➢ **任务情景**

公司财务部门及销售部门的办公网络部署完成后，两个部门之间无法直接通信。由于即将开展月底计酬业务，财务部门需要访问销售部门的业绩数据。小李打算通过增加三层交换机来实现两个部门的网络互访。

➢ **任务分析**

➢ 学会在三层交换机上划分 VLAN 并配置接口状态；
➢ 学会配置三层交换机的 VLANIF 接口；
➢ 学会查看三层交换机的工作状态。

➢ **实施准备**

1. HUAWEI S5720-36PC-IE 交换机 1 台；
2. HUAWEI S5720-28P-LI-AC 交换机 1 台；
3. PC 2 台；
4. 直通型双绞线 2 根；
5. 交叉型双绞线 1 根；
6. Console 通信线缆 1 根。

➢ **实施步骤**

1. 按照任务目标设计实训拓扑结构，并按照要求连接网络设备、互连接口。拓扑图如图 3-1-1 所示。

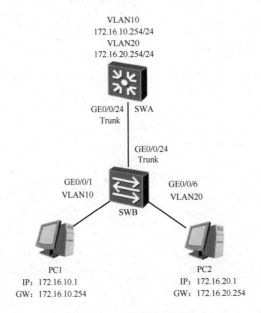

图 3-1-1　配置拓扑图

2. 按照网络功能需求规划的网络设备及 PC 的 IP 地址、子网掩码和网关，设备地址分配如表 3-1-1 所示。PC1 和 PC2 IP 地址配置分别如图 3-1-2 和图 3-1-3 所示。

表 3-1-1 设备地址分配表

设备名称	接口	IP 地址	掩码(Mask)	网关
SWA	GE0/0/24	Trunk		无
SWA	VLAN 10	172.16.10.254	255.255.255.0	无
SWA	VLAN 20	172.16.20.254	255.255.255.0	无
SWB	GE0/0/1～GE0/0/5	VLAN 10(CaiWuBu)		
SWB	GE0/0/6～GE0/0/10	VLAN 20(XiaoShouBu)		
SWB	GE0/0/24	Trunk		
PC1	Ethernet	172.16.10.1	255.255.255.0	172.16.10.254
PC2	Ethernet	172.16.20.1	255.255.255.0	172.16.20.254

图 3-1-2 PC1 IP 地址配置

图 3-1-3 PC2 IP 地址配置

3. 在交换机 SWB 上配置主机名。

```
<HUAWEI>system-view    //进入系统视图
Enter system view, return user view with Ctrl+Z.
#提示：如需返回用户视图，请按组合键<Ctrl>+<Z>
[HUAWEI] sysname SWB   //把交换机的默认设备名称HUAWEI修改为SWB
#提示：修改生效，交换机设备名称已变为SWB
```

4. 在交换机 SWB 上创建 VLAN10、VLAN20，并将交换机 GE0/0/1～GE0/0/5 接口划分到 VLAN10、GE0/0/6～GE0/0/10 接口划分到 VLAN20。

```
[SWB]vlan 10     //创建VLAN10
Info: This operation maytake a few seconds. Please wait for a moment...done.
#提示：系统提示此操作需要几秒时间，请等待完成后再继续操作
```

```
[SWB-Vlan10]description CaiWuBu            //命名VLAN10为CaiWuBu
#提示：为VLAN命名可以帮助管理员了解VLAN的属性及成员
[SWB-Vlan10]quit          //退出VLAN配置视图
[SWB]vlan 20              //创建VLAN20
[SWB-Vlan20]description XiaoShouBu         //命名VLAN20为XiaoShouBu
[SWB-Vlan20]quit          //退出VLAN配置视图
[SWB]interface range GigabitEthernet 0/0/1 to GigabitEthernet 0/0/5
 //进入接口GE0/0/1~GE0/0/5
#提示：range表示范围，此处范围指GE0/0/1~GE0/0/15共5个接口，以下简称GE0/0/x接口，输入命令行也可简写
[SWB-port-group]port link-type access//设置端口GE0/0/1~GE0/0/5的端口类型为access
[SWB-port-group]port default vlan 10  //设置端口GE0/0/1~GE0/0/5加入VLAN10
#提示：此命令行由系统自动执行，无须手动输入，以下5行情况相同
[SWB-GigabitEthernet0/0/1]port default vlan 10
[SWB-GigabitEthernet0/0/2]port default vlan 10
[SWB-GigabitEthernet0/0/3]port default vlan 10
[SWB-GigabitEthernet0/0/4]port default vlan 10
[SWB-GigabitEthernet0/0/5]port default vlan 10
[SWB]interface range GigabitEthernet 0/0/6 to GigabitEthernet 0/0/10
 //进入接口GE0/0/6~GE0/0/10
[SWB-port-group]port link-type access      //设置端口GE0/0/6~GE0/0/10的端口类型为access
 [SWB-port-group]port default vlan 20   //将端口GE0/0/6~GE0/0/10划分到VLAN20
#提示：以下命令行由系统自动执行，无须人为输入，执行结果为依次转换GE0/0/6~GE0/0/10端口加入VLAN20
[SWB-GigabitEthernet0/0/6]port default vlan 20
[SWB-GigabitEthernet0/0/7]port default vlan 20
[SWB-GigabitEthernet0/0/8]port default vlan 20
[SWB-GigabitEthernet0/0/9]port default vlan 20
[SWB-GigabitEthernet0/0/10]port default vlan 20
[SWB-port-group]quit   //退出接口配置视图
```

5. 把交换机 SWB 的 GE0/0/24 端口配置为 Trunk，允许所有 VLAN 流量通过。

```
[SWB]interface GigabitEthernet 0/0/24 //进入接口GE0/0/24配置视图
[SWB-GigabitEthernet0/0/24]port link-type trunk   //设置端口类型为Trunk
Info: This operation maytake a few seconds. Please wait for a moment...done.
 [SWB-GigabitEthernet0/0/24]port trunk allow-pass vlan 10 20 //允许VLAN10、VLAN20的流量通过Trunk类型的端口
#提示：华为交换机默认只支持VLAN1通过，必须输入此命令以允许Trunk端口放行VLAN10、VLAN20的流量
Info: This operation may take a few seconds. Please wait a moment...done.
#提示：系统提示此操作需要花费几秒时间，请等待命令执行后再继续其他操作
[SWB-GigabitEthernet0/0/24]quit   //退出接口配置视图
[SWB]quit    //退出系统视图
<SWB>save    //保存配置
The current configuration (excluding the configurations of unregistered boards or cards) will be written to flash:/vrpcfg.zip.
#提示：系统提示当前配置将会被保存到flash:/vrpcfg.zip文件中，并覆盖原有文件，要求确认
Are you sure to continue?[Y/N]y    //输入yes确认保存操作
Now saving the current configuration to the slot 0.
Apr  2 2000 00:35:41 SWB %%01CFM/4/SAVE(s)[6]:The user chose Y when deciding whether to save the configuration to the device...
```

```
Save the configuration successfully.
#提示：系统提示保存成功
```

6. 在交换机 SWA 上配置主机名。

```
<HUAWEI>system-view          //进入系统视图
Enter system view, return user view with Ctrl+Z.
[HUAWEI]sysname SWA          //设置交换机设备名称为SWA
```

7. 在交换机 SWA 上创建 VLAN10、VLAN20，并为 VLAN 命名，为 VLAN 配置 IP 地址。

```
[SWA]vlan batch 10 20     //连续创建VLAN10、VLAN20
Info: This operation maytake a few seconds. Please wait for a moment...done.
[SWA]interface Vlanif 10 //开启VLANIF接口
#提示：配置VLANIF10接口的目的是使财务部门的VLAN能够访问销售部门的VLAN
Dec 11 2018 14:47:46 SWA %%01IFNET/4/IF_STATE(l)[0]:Interface Vlanif10 has turned into
UP state.
    [SWA-Vlanif10]description CaiWuBu //将VLAN命名为CaiWuBu
    [SWA-Vlanif10]ip address 172.16.10.254 24   //设置VLANIF接口的IP地址为172.16.10.254、掩
码为255.255.255.0
    #提示：此处配置VLANIF接口IP地址后，交换机上虚拟接口VLANIF10就承担了VLAN10的网关，VLAN10的流量
即可转发到VLAN20
    Dec 11 2018 14:47:59 SWA %%01IFNET/4/LINK_STATE(l)[1]:The line protocol IP on the
interface Vlanif10 has entered the UP state.
    #提示：当系统配置了VLAN接口后，协议状态为UP，表示此虚拟接口（VLANIF）已经生效
    [SWA-Vlanif10]quit    //退出VLAN10虚拟接口配置视图
    [SWA]interface Vlanif 20  //配置VLANIF20接口，使销售部门的VLAN能够访问财务部门的VLAN
    Dec 11 2018 14:50:14 SWA %%01IFNET/4/IF_STATE(l)[3]:Interface Vlanif20 has turned into
UP state.
    #提示：当系统启用了VLAN接口后，协议状态为UP，表示此虚拟接口（VLANIF）已经进入工作状态
    [SWA-Vlanif20]description XiaoShouBu    //命名VLAN20为XiaoShouBu，与SWB保持一致
    [SWA-Vlanif20]ip address 172.16.20.254 24   //设置VLANIF20虚拟接口的IP地址为172.6.20.254、
掩码为255.255.255.0
    Dec 11 2018 14:50:29 SWA %%01IFNET/4/LINK_STATE(l)[4]:The line protocol IP on the
interface VLANIF20 has entered the UP state.
    #提示：此虚拟接口（VLANIF20）已经生效
    [SWA-Vlanif20]quit   //退出VLANIF虚拟接口配置视图
```

小贴士

VLAN10 和 VLAN20 的用户是通过 SWA 上的 VLANIF 端口互通的。VLANIF 端口是交换机上的虚拟端口，可做为某个 VLAN 的逻辑网关，转发 VLAN 流量到其他网络。

8. 配置交换机 SWA 与 SWB 的互连接口 GE0/0/24 的端口类型为 Trunk，允许所有 VLAN 流量通过。

```
[SWA]interface GigabitEthernet 0/0/24    //进入接口GE0/0/24
[SWA-GigabitEthernet0/0/24]port link-type trunk    //设置端口为Trunk模式
Info: This operation may takea few seconds. Please wait for a moment...done.
#提示：系统提示此操作需要花费几秒时间，请等待
```

```
    Dec 11 2018 14:52:55 SWA %%01IFNET/4/LINK_STATE(l)[6]:The line protocol IP on the
interface Vlanif10 has entered the DOWN state.
    #提示：VLANIF10虚拟接口的状态转为DOWN，即接口关闭
    Dec 11 2018 14:52:55 SWA %%01IFNET/4/IF_STATE(l)[7]:Interface Vlanif20 has turned into
DOWN state.
    #提示：VLANIF20虚拟接口的状态转为DOWN，即接口关闭
    [SWA-GigabitEthernet0/0/24]port trunk allow-pass vlan 10 20   //华为交换机默认不支持其他
VLAN通过，VLAN1除外,所以要允许Trunk端口通过VLAN10、VLAN20
    Info: This operation may take a few seconds. Please wait a moment...done.
    #提示：系统提示此操作需要花费几秒时间，请等待
    Dec 11 2018 14:54:21 SWA %%01IFNET/4/LINK_STATE(l)[8]:The line protocol IP on the
interface Vlanif10 has entered the UP state.
    #提示：VLANIF10虚拟接口的状态转为UP，即接口已打开
    Dec 11 2018 14:54:21 SWA %%01IFNET/4/LINK_STATE(l)[9]:The line protocol IP on the
interface Vlanif20 has entered the UP state.
    #提示：VLANIF20虚拟接口的状态转为UP，即接口已打开
    [SWA-GigabitEthernet0/0/24] quit   //退出接口配置视图，返回系统视图
    [SWA]quit      //退出系统视图
    <SWA>save      //返回用户视图，并保存当前配置
```

9. 配置完成，测试网络连通性。

使用 ping 命令从 PC1 向 PC2 发送 4 个测试数据包，收到 PC2 的应答信息，连通性测试结果如图 3-1-4 所示，说明财务部门与销售部门已经实现部门之间的网络互通。

图 3-1-4　连通性测试结果

任务总结与思考

本任务重点讲述使用交换机的 VLANIF 接口实现不同部门网络互访的配置方法，同时介绍交换机的各种接口状态的特点。

思考以下两个问题。

1．交换机的物理接口与逻辑端口有什么不同？各有什么特点？

2．本任务中用到了交换机的几种视图？如何切换？

 知识补给

交换机通过 VLANIF 接口之间相互转发流量，可以实现 VLAN 之间的互访。通常，HUAWEIS2750EI 以前的版本只支持配置一个 VLANIF 接口，这意味着仅靠该交换机无法实现不同 VLAN 间的互访。

HUAWEIS2750EI、HUAWEIS5700S-LI、HUAWEIS5700LI、HUAWEIS5710-X-LI 交换机均有 8 个 VLANIF 接口，可以实现 8 个 VLAN 的三层互访，通常称为三层交换机。

➢ 命令格式。

　　interface vlanif vlan-id

　　undo interface vlanif vlan-id

➢ 命令功能。

　　interface vlanif 命令用来创建 VLANIF 接口，并进入 VLANIF 接口视图。

　　undo interface vlanif 命令用来删除 VLANIF 接口。

　　默认情况下，VLANIF 接口没有被创建。

➢ 使用实例。

　　创建 VLANIF2 接口，并进入 VLANIF2 接口视图。

　　<HUAWEI>system-view

　　[HUAWEI]vlan 2

　　[HUAWEI-vlan2] quit

　　[HUAWEI]interface vlanif 2

➢ 注意事项。

　　（1）创建某 VLAN 对应的 VLANIF 接口后，该 VLAN 不能再用作子接口配置的 VLAN。

　　（2）如果 VLANIF 接口已经存在，则 interface vlanif 命令只用来进入 VLANIF 接口视图。

　　（3）当 VLANIF 接口作为 Telnet 连接设备的管理 VLANIF 时，用户接入 VLAN 的编号和管理 VLAN 的编号不能共用，否则可能会导致设备 Telnet 无法连接。

3.2 安装路由器

➢ 任务情景

宇信公司根据网上业务迁移需要,购置了一批HUAWEIAR2220E路由器。在路由器接入网络之前,小李需要先了解这批设备的外观、安装、连接方法,然后把路由器安装到机柜中。

➢ 任务分析

- ➢ 学会使用安装工具;
- ➢ 了解路由器的安装规范;
- ➢ 学会安装并连接路由器。

➢ 实施准备

1. HUAWEIAR2220E 路由器 1 台;
2. PC 1 台;
3. 交叉型双绞线 1 根(长度为 2m);
4. 路由器原厂 Console 通信电缆 1 根;
5. 设备安装工具箱。

【提示:设备安装工具箱包括十字旋具、一字旋具、浮动螺母、挂耳、M4 螺钉、M6 螺钉。其中,挂耳包含在随路由器发货的安装附件包中,其余工具都需要用户单独购买。】

➢ 实施步骤

1. 安装空间检查。检查机柜固定是否牢固,并在机柜内为路由器预留安装空间。HUAWEIAR2220E 的摆放高度为 1U,对应机柜两侧 3 个方孔的位置,要求路由器四周均留出至少 50mm 的距离,以用于散热。

【提示:1U 即 1UNIT(单元),是一种表示网络设备外部尺寸的单位,1U=4.445cm。】

2. 安装条件检查。对照路由器的随机文档《硬件描述》的要求,检查机柜内的供电电源是否工作正常,电压是否满足路由器工作要求,以及通信线缆、安装工具、防护工具是否到位。

【提示:AR 系列路由器可正常工作的电压范围请参见对应产品的随机文档《硬件描述》。】

3. 安装路由器挂耳。使用十字旋具,利用 M4 螺钉将挂耳固定在路由器两侧,如图 3-2-1 所示。

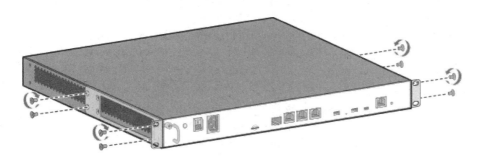

图 3-2-1　安装路由器挂耳

【提示：挂耳可以安装在靠近路由器前面板的两侧，也可以安装在靠近路由器后面板的两侧。】

4．标记路由器安装位置。确定路由器安装在机柜内第几个单元（UNIT）的高度，用记号笔在机柜上标记浮动螺母和 L 形滑道底边位置，如图 3-2-2 所示。

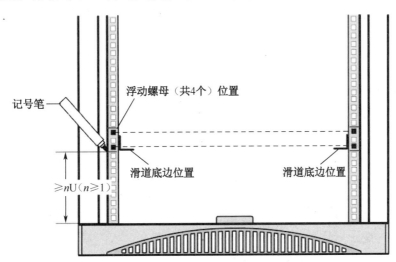

图 3-2-2　标记路由器安装位置

5．安装浮动螺母。根据标记好的浮动螺母位置，保证左右对应的浮动螺母在同一个水平面上，使用一字旋具在机柜前方孔条上安装 4 个浮动螺母，左右各 2 个，如图 3-3-3 所示。

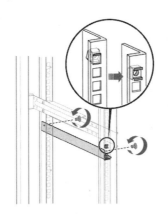

图 3-2-3　安装浮动螺母

6. 安装 L 形滑道。将 L 形滑道的底边位置与记号笔标记的下边缘位置对齐，利用 M5 螺钉将 L 形滑道固定在机柜的左右两侧，安装位置如图 3-2-4 所示。

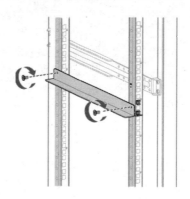

图 3-2-4　安装 L 形滑道

7. 将路由器安装到机柜。将路由器搬到机柜中，放置在 L 形滑道上，平稳地推入机柜中，直至挂耳与机柜前方孔条表面贴紧为止，利用 M6 螺钉将挂耳固定到机柜上，步骤及效果如图 3-2-5 中步骤ⓐ～步骤ⓑ所示。

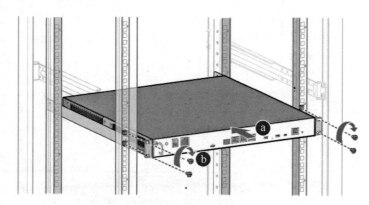

图 3-2-5　安装路由器到机柜

8. 认识接口。首先从背板位置观察路由器的外观和接口类型。路由器外观如图 3-2-6 所示。

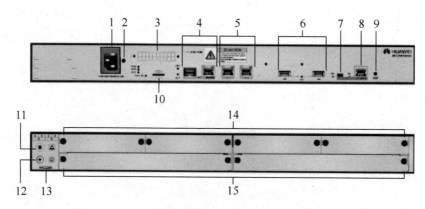

图 3-2-6　HUAWEIAR2220E 路由器外观

9. 安装接地线缆。使用十字旋具，拧下位于后面板接口 12 的 M4 螺钉，将接地线缆的 M4 端对准接地端子上的螺钉孔，用 M4 螺钉固定，将接地线缆的 M6 端与机柜的接地端相连，分解步骤如图 3-3-7 中步骤ⓐ～步骤ⓒ所示。

【提示：进行安装操作时必须佩戴防静电腕带，并确保防静电腕带的一端接地，另一端与佩戴者的皮肤良好接触。】

图 3-2-7　连接路由器地线

 小贴士

在路由器包装箱内找到原厂接地线缆。完成连接后需要检查接地线缆与接地端子的连接是否牢固可靠，并要求路由器接地点与接地端之间的电阻小于 5Ω。

10. 连接电源线。路由器正面的接口 1 为电源接口（见图 3-3-6），选择随厂配备的电源线进行连接，接入 220V 交流电源后，打开电源开关，路由器上电后启动，路由器电源指示灯 PWR 呈绿色常亮状态。

【提示：路由器包装内的电源线作为随设备发货附件之一，只可与本包装内的主机配套使用，不可用于其他设备。】

小贴士

如果外部供电系统提供的是交流制式插座，则应配套使用相应制式的交流电源线缆；如果外部供电系统提供的是直流配电盒，则应配套使用直流电源线缆。

任务总结与思考

本任务重点讲述了路由器的外观、接口及安装到机柜的方法，使学生对路由器及其他网络设备有了初步的认识。

思考以下两个问题。

1. 初次认识一种网络设备时，怎样才能更快速地了解其性能及参数？
2. 尝试说出路由器与交换机在外观、功能或应用等方面的区别。

知识补给

3.2.1 HUAWEIAR2220E 路由器的接口与槽位

HUAWEIAR2220E 路由器的接口与槽位排列如图 3-2-6 所示，相应功能说明如表 3-2-1 所示。

表 3-2-1　路由器的接口与槽位功能说明

序号	接口名称	功能说明
1	交流电源线接口	通过交流电源线缆将设备连接到外部电源
2	电源线防松脱卡扣安装孔	用来绑定电源线，防止电源线松脱
3	RPS 电源线接口	通过 RPS150 供电通信线缆连接 150W RPS 电源模块
4	GE 管理口（默认 IP 地址为 192.168.1.1/24）	千光以太网（GigEthernet）复用接口，左侧为光纤接口，右侧为 UTP 电缆接口，二者不能同时使用
5	WAN 侧接口	2 个千光以太网（GigEthernet）接口，适用于 UTP 电缆连接
6	USB 接口	用于连接 3G 的 USB MODEM，上面两个小孔用于安装 USB 防护套
7	MiniUSB 接口	MiniUSB 接口和 Console 接口不能同时使用
8	Console 接口	路由器的管理接口
9	RST 按钮	复位按钮，用于手动复位设备（复位设备会导致业务中断，需慎用）
10	Micro SD 卡插槽	支持外接 Micro SD 卡
11	ESD 插孔	对设备进行维护操作时，需要佩戴防静电腕带，防静电腕带的一端要插在 ESD 插孔里
12	接地点	使用接地线缆将设备接地，以防雷、防干扰
13	产品型号丝印	产品型号标签
14	4 个 SIC 槽位	用于安装多功能串行接口卡
15	2 个 WSIC 槽位	用于安装语音模块接口卡

3.2.2 HUAWEIAR2220E 路由器的指示灯

HUAWEIAR2220E 路由器的指示灯较多，具体位置与标识如图 3-2-8 所示，其具体含义及状态说明如表 3-2-2 所示。

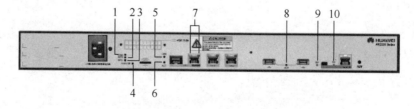

图 3-2-8　HUAWEIAR2220E 路由器指示灯位置与标识

表 3-2-2　HUAWEIAR2220EAR2220E 指示灯含义及状态说明

序号	指示灯名称	颜色	含义及状态说明
1	PWR（电源指示灯）	绿色	设备内部电源供电正常
		红色	设备内部电源供电异常
		常灭	设备未上电
2	SYS（系统指示灯）	绿色	慢闪：系统处于正常运行状态
			快闪：系统处于上电加载或复位启动状态
		红色	常亮：单板有影响业务且无法自动恢复的故障，需要人工干预
		常灭	红灯、绿灯均不亮，表明软件未运行或处于复位状态
3	RPS（电源指示）	绿色	常亮：RPS 电源在位
		黄色	常亮：RPS 电源已连接但状态异常
			闪烁：RPS 电源正在给设备供电
		常灭	RPS 电源未连接
4	Micro SD 卡指示灯	绿色	常亮：链路已经连通
			闪烁：有数据收发
			常灭：无卡
5 和 6	GE 光接口指示灯： 5 为 LINK 6 为 ACT	LINK 灯	LINK 灯常亮：链路已经连通
			LINK 灯常灭：链路无链接
		ACT 灯	ACT 灯闪烁：有数据收/发
			ACT 灯常灭：无数据收/发
7	GE 电接口指示灯	绿色	LINK 灯常亮：链路已经连通
			LINK 灯常灭：链路无链接
		黄色	ACT 灯闪烁：有数据收/发
			ACT 灯常灭：无数据收/发
8	ACT（USB）	红绿双色	绿色常亮：U 盘引导部署正确完成
			绿色闪烁：U 盘引导部署正在进行中
			红色常亮：U 盘引导部署失败
			常灭：未插开局 U 盘，USB 接口故障或指示灯故障
9	EN（MiniUSB 接口）	绿色	常亮：当前是 MiniUSB 接口使能
			常灭：当前不是 MiniUSB 接口使能
10	EN（CON/AUX 接口）	绿色	常亮：当前是 CON/AUX 接口使能
			常灭：当前不是 CON/AUX 接口使能

【提示：CON/AUX 接口和 MiniUSB 接口是复用的，同一时刻只有一个可以使用。默认 CON/AUX 接口有效，对应的 EN 指示灯绿色常亮，无论是否插线缆。】

任务拓展

1．认识接口单板。

路由器安装完成后，我们已经熟悉了路由器的接口及槽位，接下来了解一下在路由器背板的槽位上可以安装什么样的接口模块，并尝试安装这些接口模块。路由器的接口往往是由一块电子元器件板卡承载的，通常称为单板电路，简称单板。单板电路上布满了各种功能芯片，并对外提供指示灯、按钮、接口，还提供便于安装的拉手或扣板。

本任务提到的 HUAWEIAR2220E 路由器背板上有 4 个 SIC 槽位和 2 个 WSIC 槽位，可以安装对应的单板模块来增加路由器的串行接口，以实现广域网的接入。支持的典型 SIC 单板一般包含 1～2 个串行接口，如图 3-2-9 所示。

图 3-2-9　典型 SIC 单板

HUAWEIAR2220E 路由器支持的典型 WSIC 单板的外观如图 3-2-10 所示，一般包含多个异步语音接口。

图 3-2-10　典型的 WSIC 单板外观

2．安装接口单板。

本任务中路由器背板上暂时安装的是假面板。如果需要扩展路由器的广域网接口，就需要将接口单板安装到路由器的对应槽位中，下面介绍如何安装单板。

准备工具和附件，包括防静电腕带和十字旋具等。

操作步骤

（1）佩戴防静电腕带，并确保防静电腕带的一端接地，另一端与佩戴者的皮肤良好接触。

（2）拧开槽位上假面板的松不脱螺钉，并取下假面板。

（3）将单板上的扳手打开，沿插槽的导轨水平推入单板，直至单板的面板贴紧设备，如图 3-2-11 中的步骤ⓐ所示。

（4）将扳手往里压紧，直至单板完全进入插槽，并拧紧单板两侧的松不脱螺钉，如图 3-2-11 中的步骤ⓑ所示。

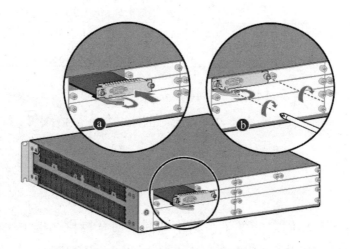

图 3-2-11　安装路由器单板

【提示：单板必须安装在支持该单板的路由器槽位中，单板与路由器的配套关系请参见对应硬件产品随机文档《硬件描述》中的"单板"部分的内容。】

小技巧

安装单板时需要缓慢插入，如果插入过程中遇到较大阻力或单板位置出现偏斜，则必须将单板拔出后重新插入，严禁强行安装，以免损坏单板和路由器背板的连接器。

3.3 管理路由器

➢ **任务情景**

宇信公司采购了一批 HUAWEIAR2220E 路由器，准备将其作为该企业网络出口设备。在将设备接入网络前，小李需首先通过控制台对新出厂的路由器做带外管理，并要了解设备的硬件和软件特性。

➢ **任务分析**

- ➢ 学会做路由器的带外管理；
- ➢ 了解路由器的硬件信息；
- ➢ 了解路由器的文件系统。

➢ **实施准备**

1. HUAWEIAR2220E 路由器 1 台；
2. PC 1 台；
3. 直通型双绞线 1 根；
4. Console 通信线缆 1 根；
5. 终端仿真软件（本书以 Secure CRT 为例）。

小贴士

路由器或交换机在接入网络之前，只能通过控制台（Console）接口进行现场配置或管理，称为带外管理；接入网络之后，系统为网络接口配置了 IP 地址，与其他设备连通后，可以通过网络进行远程管理，称为带内管理。

➢ **实施步骤**

1. 使用 Console 通信线缆将路由器的 Console 接口与 PC 的 COM 端口进行物理连接。

管理路由器拓扑图如图 3-3-1 所示。

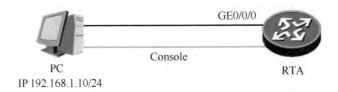

图 3-3-1　管理路由器拓扑图

2．在 PC 上右击"我的电脑"，打开"计算机管理"窗口，查看连接路由器的 COM 端口编号，如图 3-3-2 所示。

图 3-3-2　查看连接路由器的 COM 端口编号

3．在资源管理器中打开 Secure CRT 软件所在文件夹，再打开应用程序窗口，双击可执行程序，如图 3-3-3 所示。

图 3-3-3　查找可执行程序

4. 在打开的窗口中单击"新建"按钮，新建一个 CRT 连接，如图 3-3-4 所示。

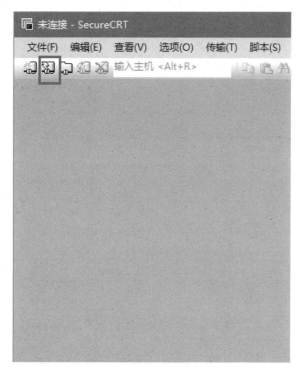

图 3-3-4　"新建"按钮

5. 终端软件的通信参数设置需与设备的默认值保持一致，分别设置传输速率（波特率）为 9600bit/s、数据位为 8 位、停止位为 1 位，并设置"奇偶检验"为"None"，不勾选"流控"中的选项。单击"连接"按钮，完成设置，如图 3-3-5 所示。

图 3-3-5　设置连接的接口及通信参数

6. 连接成功后，系统要求输入路由器的默认用户名和密码，如图 3-3-6 所示。

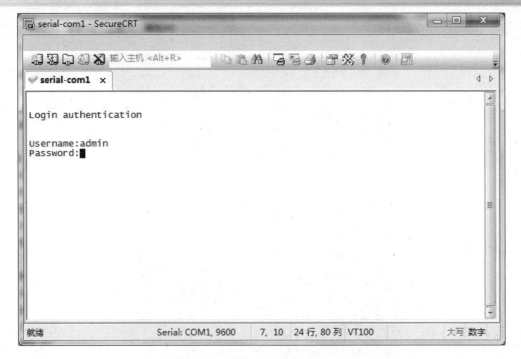

图 3-3-6　输入默认用户名和密码

【提示：华为路由器默认用户名为 admin，默认密码为 Admin@huawei。】

7. 验证成功后，进入系统初始界面，此时用户可以输入命令，对路由器进行配置。当用户需要帮助时可以随时输入?以寻求系统提示。路由器登录界面如图 3-3-7 所示。

```
------------------------------------------------------------
User last login information:
------------------------------------------------------------
Access Type: Serial
IP-Address : --
Time       : 2019-01-19 20:20:20+00:00
------------------------------------------------------------
Warning: There is a risk in the current configuration file. Please save configuration as soon as possible.
<Huawei>system-view
```

图 3-3-7　路由器登录界面

进入路由器系统视图，修改主机名为 Router1。

```
<Huawei>system-view     //进入系统视图
Enter system view, return user view with Ctrl+Z.   //系统提示进入系统视图，退出请按组合键<Ctrl>+<Z>
[Huawei]sysname Router1    //注意，提示符由<>变为[]，修改主机名为Router1
```

8. 为防止非授权登录，修改系统密码。

```
[Router1]aaa            //进入aaa账户模式
[Router1-aaa]
[Router1-aaa]local-user admin password irreversible-cipher HNXXGCXX@student
//修改admin用户密码为HNXXGCXX@student
Please enter old password:       //系统要求输入admin用户原密码
Info: The password is changed successfully.   //密码已修改成功
```

```
Info: After you change the rights (including the password, access type, FTP directory,
and level) of a local user, the rights of users already online do not change. The change takes
effect to users who go online after the change.
```

9. 查看路由器系统信息。

```
[Router1]display version        //查看当前系统版本号
Huawei Versatile Routing Platform Software
VRP (R) software, Version 5.160 (AR2200 V200R007C00SPC900)      //当前软件版本号为5.160
Copyright (C) 2011-2016 HUAWEI TECH CO., LTD                    //版本发行时间
Huawei AR2220E Router uptime is 0 week, 0 day, 0 hour, 25 minutes//路由器上电时间
BKP 0 version information:                                      //背板信息
1. PCB        Version  : AR01BAK2C VER.B      //PCB版本信息
2. If Supporting PoE : No                     //是否支持以太网充电(PoE)功能
3. Board      Type     : AR2220E              //主板类型
4. MPU Slot Quantity : 1                      //主控板插槽数目
5. LPU Slot Quantity : 6                      //业务板插槽数目
...                                           //省略部分硬件信息
```

10. 查看系统文件信息。

```
<Router1>dir        //查看当前路由器文件
Directory of flash:/               //flash:/ 为路由器的系统文件主目录

  Idx  Attr     Size(Byte)  Date        Time(LMT)   FileName
   0   -rw-         27,554  Jan 19 2019 20:13:21    mon_file.txt
   1   -rw-            120  Dec 24 2018 16:35:46    vrpcfg.zip  //系统配置文件
   2-  rw-    126,538,240  Feb 28 2018 16:07:02    AR2220E-V200R007C00SPC900.cc  // 操作系统
文件
   3   -rw-              0  Mar 06 2018 02:05:10    brdxpon_snmp_cfg.efs
   4   drw-             -   Mar 06 2018 02:05:34    update
   5   drw-             -   Mar 06 2018 02:05:56    shelldir
   6   drw-             -   Jan 19 2019 20:39:37    localuser
   7   drw-             -   Mar 06 2018 02:06:10    dhcp
   8   -rw-            783  Dec 06 2018 11:14:58    default_local.cer
   9-  rw-        5,899,648 Mar 07 2018 15:57:56    AR2220E-V200R007SPH013.pat
  10   -rw-            430  Dec 24 2018 16:31:30    private-data.txt
  11   -rw-          2,240  Jan 19 2019 20:14:10    mon_lpu_file.txt
510,484 KB total available (243,272 KB free)
<Router1>
```
#提示：Idx 代表序号，-rw-代表读/写权限，d代表目录，Size代表文件大小，Date代表创建日期，FileName代表文件名

11. 简要查看路由器接口情况。

```
[Router1]display ip interfacebrief//简要查看当前接口的IP地址配置及状态
...                               //省略部分介绍性信息

Interface                    IP Address/Mask        Physical    Protocol
Cellular0/0/0                unassigned             down        down
Cellular0/0/1                unassigned             down        down
GigabitEthernet0/0/0         192.168.1.1/24         up          up
GigabitEthernet0/0/1         unassigned             down        down
GigabitEthernet0/0/2         unassigned             down        down
GigabitEthernet0/0/3         unassigned             up          down
```

```
  NULL0                    unassigned           up      up(s)

[Router1]quit
```
#提示：以上显示了接口的类型、ID号、IP地址/掩码、物理状态及协议状态，从中可以看出，新出厂的路由器的GE0/0/0接口配置了默认地址192.168.1.1/24

 任务总结与思考

本任务重点讲述了在利用 Console 接口连接 PC 后如何对出厂路由器进行初始配置。

思考以下两个问题。

1．在本任务拓扑设计中，PC 与路由器之间为什么要通过两根线缆连接？它们分别起到什么作用？

2．观察路由器的接口，阅读路由器说明书，管理路由器除通过 Console 接口方式外，还可以通过哪些方式进行管理？

 任务拓展

华为路由器内置了 FTP 服务器功能，但出厂时默认是关闭此项服务的，下面介绍启用此功能的操作步骤。

1．进入路由器系统视图，启用 FTP 服务功能。
```
<Router1>system-view              //进入系统视图
Enter system view, return user view with Ctrl+Z.
[Router1]ftp server enable        //开启FTP服务器功能
Info: Succeeded in starting the FTP server     //系统提示FTP服务器启动成功
Info: FTP is insecure, recommended to use SFTP with encryption features.
```

2．创建 FTP 用户。
```
[Router1]aaa                                      //进入aaa用户管理模式
[Router1-aaa]local-user ftp password cipher ftp123456
//创建本地用户，用户名为 ftp 密码为ftp123456
Info: Add a new user.     //系统提示新增一个用户
```

3．设置 FTP 用户权限及属性。
```
[Router1-aaa]local-user ftp service-type  ftp  //登录本地FTP用户，尝试FTP服务
Info: The cipher password has been changed to an irreversible-cipher password.
Warning: The user access modes include Telnet, FTP or HTTP, and so security risks exist.
Info: After you change the rights (including the password, access type, FTP directory,
and level) of a local user, the rights of users already online do not change. The change takes
effect to users who go online after the change.
```
#提示：系统提示此用户还可进行Telnet、FTP、HTTP访问，但存在风险。如果尝试改变本地用户的权限，则对已经登录的用户无效，只对修改权限后登录的用户有效
```
[Router1-aaa]local-user ftp ftp-directory flash:/   //指定FTP用户可访问的目录
```
#提示：如果不配置FTP用户可访问的目录，则FTP用户无法登录设备
```
Info: After you change the rights (including the password, access type, FTP directory,
and level) of a local user, the rights of users already online do not change. The change takes
```

```
effect to users who go online after the change.
    [Router1-aaa]local-user ftp access-limit 10   //设置FTP最大连接数为10
    #提示：此操作表示此FTP服务器最多只允许10个本地用户同时登录
    [Router1-aaa]local-user ftp privilege level 15   //设置用户等级为15级
    #提示：15即为最高用户级别
    Info: After you change the rights (including the password, access type, FTP directory,
and level) of a local user, the rights of users already online do not change. The change takes
effect to users who go online after the change.
    [Router1-aaa]quit              //退出aaa设置模式
```

4．设置 PC 的 IP 地址为 192.168.1.10/24，打开 CMD 窗口，连接路由器 FTP 服务，输入新设置的用户名密码，如图 3-3-8 所示。

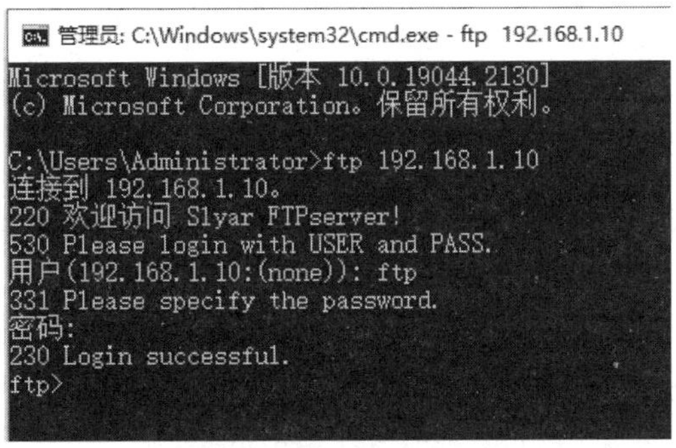

```
管理员: C:\Windows\system32\cmd.exe - ftp  192.168.1.10
Microsoft Windows [版本 10.0.19044.2130]
(c) Microsoft Corporation。保留所有权利。

C:\Users\Administrator>ftp 192.168.1.10
连接到 192.168.1.10。
220 欢迎访问 Slyar FTPserver!
530 Please login with USER and PASS.
用户(192.168.1.10:(none)): ftp
331 Please specify the password.
密码:
230 Login successful.
ftp>
```

图 3-3-8　登录 FTP 服务

5．浏览 FTP 服务器文件系统，并成功下载 mon_file.txt 的文件，如图 3-3-9 所示。

【提示：图 3-3-9 中用到了 FTP 应用程序的命令解释，dir 为列目录，get 为下载文件，put 为上传文件。】

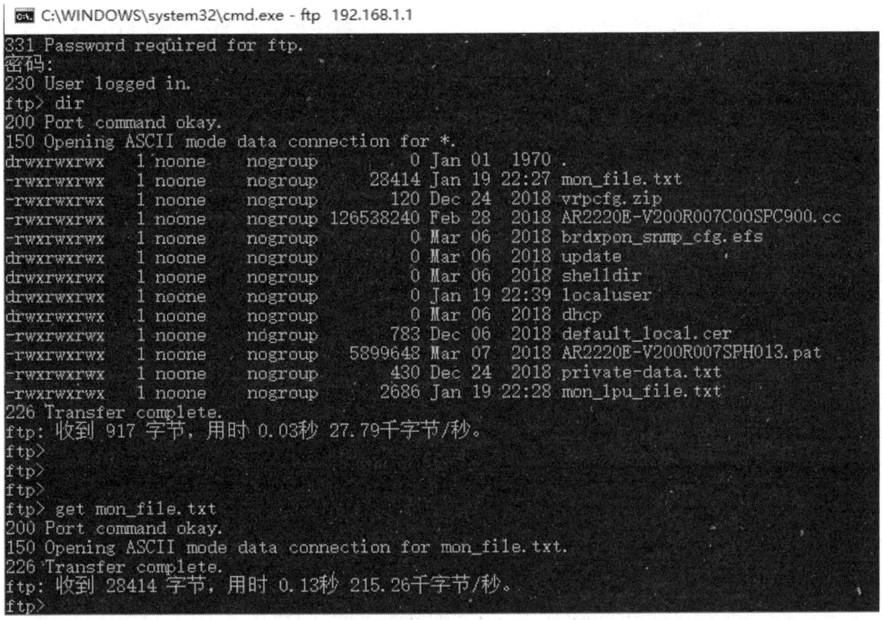

```
C:\WINDOWS\system32\cmd.exe - ftp  192.168.1.1
331 Password required for ftp.
密码:
230 User logged in.
ftp> dir
200 Port command okay.
150 Opening ASCII mode data connection for *.
drwxrwxrwx   1 noone    nogroup          0 Jan 01  1970 .
-rwxrwxrwx   1 noone    nogroup      28414 Jan 19 22:27 mon_file.txt
-rwxrwxrwx   1 noone    nogroup        120 Dec 24  2018 vrpcfg.zip
-rwxrwxrwx   1 noone    nogroup  126538240 Feb 28  2018 AR2220E-V200R007C00SPC900.cc
-rwxrwxrwx   1 noone    nogroup          0 Mar 06  2018 brdxpon_snmp_cfg.efs
drwxrwxrwx   1 noone    nogroup          0 Mar 06  2018 update
drwxrwxrwx   1 noone    nogroup          0 Mar 06  2018 shelldir
drwxrwxrwx   1 noone    nogroup          0 Jan 19 22:39 localuser
drwxrwxrwx   1 noone    nogroup          0 Mar 06  2018 dhcp
-rwxrwxrwx   1 noone    nogroup        783 Dec 06  2018 default_local.cer
-rwxrwxrwx   1 noone    nogroup    5899648 Mar 07  2018 AR2220E-V200R007SPH013.pat
-rwxrwxrwx   1 noone    nogroup        430 Dec 24  2018 private-data.txt
-rwxrwxrwx   1 noone    nogroup       2686 Jan 19 22:28 mon_lpu_file.txt
226 Transfer complete.
ftp: 收到 917 字节，用时 0.03秒 27.79千字节/秒。
ftp>
ftp>
ftp>
ftp> get mon_file.txt
200 Port command okay.
150 Opening ASCII mode data connection for mon_file.txt.
226 Transfer complete.
ftp: 收到 28414 字节，用时 0.13秒 215.26千字节/秒。
ftp>
```

图 3-3-9　浏览目录并下载文件

> 小贴士

FTP 是一种基于传输层 TCP 端口 20 和端口 21 建立的连接，可用来进行可靠的文件传输。在本任务中将路由器配置为 FTP 服务器，采用了出厂时的默认地址 192.168.1.1，PC 作为与其直连的 FTP 客户端，只有配置与服务器同网段的 IP 地址 192.168.1.X，才能保证其连通性。

3.4 实现不同业务网络互访

实现不同业务网络互访

➢ 任务情景

公司财务部门与销售部门的办公网络部署完成了，但两个部门无法直接通信。月底计酬时财务部门需要访问销售部门的部分数据，因此小李打算通过增加三层设备路由器，利用单臂路由功能实现两个部门之间的网络互访。

➢ 任务分析

- ➢ 了解路由器的基本功能；
- ➢ 学会配置单臂路由；
- ➢ 了解路由表条目。

➢ 实施准备

1. HUAWEI AR2220 路由器 1 台；

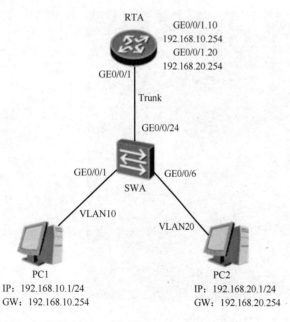

2. HUAWEI S5720-28P-LI-AC 交换机 1 台；
3. PC 2 台；
4. 直通型网线 3 根；
5. Console 通信线缆 1 根。

➢ 实施步骤

1. 按任务要求连接网络设备。单臂路由配置拓扑图如图 3-4-1 所示。

2. 按任务情景规划路由器 RTA、RTB、交换机 SWA 的 IP 地址、子网掩码和网关，设备地址配置如表 3-4-1 所示。PC 配置如图 3-4-2 和图 3-4-3 所示。

图 3-4-1 单臂路由配置拓扑图

表 3-4-1 设备地址配置

设备名称	接口	IP 地址	掩码（Mask）	网关
RTA	GE0/0/1.10	192.168.10.254	255.255.255.0	无
RTA	GE0/0/1.20	192.168.20.254	255.255.255.0	无
SWA	GE0/0/1-5	Access VLAN10		
SWA	GE0/0/6-10	Access VLAN20		
SWA	GE0/0/24	Trunk		
PC1	Ethernet	192.168.10.1	255.255.255.0	192.168.10.254
PC2	Ethernet	192.168.20.1	255.255.255.0	192.168.20.254

图 3-4-2　PC1 地址配置　　　　　　　图 3-4-3　PC2 地址配置

3．交换机启动后，配置其设备名称为 SWA。

```
<HUAWEI>system-view          //进入系统视图
#提示：交换机启动后，首先进入用户模式，系统显示<HUAWEI>,表示默认主机名为HUAWEI。在此视图中只能查看一些机器状态，不能进行配置
Enter system view, return user view with Ctrl+Z. //如需返回用户视图，则需同时按<Ctrl>+<Z>键
[HUAWEI]sysname SWA         //配置交换机设备名称为SWA
[SWA]                       //交换机的提示符变为[SWA],表示设备名称已生效
```

4．在交换机 SWA 上创建 VLAN10、VLAN20，并将 GE0/0/1～GE0/0/5 加入 VLAN10，GE0/0/6～GE0/0/10 加入 VLAN20。

```
[SWA]vlan batch 10 20       //连续创建VLAN10、VLAN20
#提示：vlan batch命令可以批量创建VLAN,一次最多输入10个不连接的VLAN号
Info: This operation may take a few seconds. Please wait for a moment...
done. //系统提示需要等待几秒时间
[SWA]interface Vlanif 10    //进入VLAN10配置视图
[SWA-Vlanif10]description CaiWuBu   //命名VLAN10为CaiWuBu
#提示：给VLAN命名主要是便于管理员了解此VLAN的属性
```

```
[SWA-Vlanif10]quit    //退出VLAN10配置视图
[SWA]interface Vlanif 20      //进入VLAN20配置视图
[SWA-Vlanif20]description XiaoShouBu   //命名VLAN20为XiaoShouBu
[SWA-Vlanif20]quit         //退出VLAN10配置视图
[SWA]interface range GigabitEthernet 0/0/1 to GigabitEthernet 0/0/5      //进入SEA交换机GE0/0/1～GE0/0/5接口
#提示：range表示范围，此处范围指GE0/0/1～GE0/0/5共5个接口
[SWA-port-group]port link-type access    //设置接口GE0/0/1～GE0/0/5的端口类型为Access
#提示：交换机与PC相连的接口只允许通过某一VLAN的流量，一般配置为Access端口类型即可满足需求
[SWA-GigabitEthernet0/0/1]port link-type access    // GE0/0/1接口的端口模式设置为Access
Info: This operation may take a few seconds. Please wait for a moment...done.
[SWA-GigabitEthernet0/0/2]port link-type access    // GE0/0/2接口的端口模式设置为access
Info: This operation may take a few seconds. Please wait for a moment...done.
[SWA-GigabitEthernet0/0/3]port link-type access    // GE0/0/3接口的端口模式设置为access
Info: This operation may take a few seconds. Please wait for a moment...done.
[SWA-GigabitEthernet0/0/4]port link-type access    // GE0/0/4接口的端口模式设置为access
Info: This operation may take a few seconds. Please wait for a moment...done.
[SWA-GigabitEthernet0/0/5]port link-type access    // GE0/0/5接口的端口模式设置为access
Info: This operation may take a few seconds. Please wait for a moment...done.
[SWA-port-group]port default VLAN 10    //将GE0/0/1～GE0/0/5接口划分到VLAN10
[SWA-GigabitEthernet0/0/1]port default VLAN 10
[SWA-GigabitEthernet0/0/2]port default VLAN 10
[SWA-GigabitEthernet0/0/3]port default VLAN 10
[SWA-GigabitEthernet0/0/4]port default VLAN 10
[SWA-GigabitEthernet0/0/5]port default VLAN 10
[SWA-port-group]quit  //退出GE0/0/1～GE0/0/5端口配置视图
[SWA]interface range GigabitEthernet 0/0/6 to GigabitEthernet 0/0/10     //进入GE0/0/6～GE0/0/10接口配置视图
#提示：range表示范围，此处范围指GE0/0/6～GE0/0/10共5个接口
[SWA-port-group]port link-type access    //设置GE0/0/6～GE0/0/10这5个接口的端口类型为Access
[SWA-GigabitEthernet0/0/6]port link-type access
Info: This operation may take a few seconds. Please wait for a moment...done.
[SWA-GigabitEthernet0/0/7]port link-type access
Info: This operation may take a few seconds. Please wait for a moment...done.
[SWA-GigabitEthernet0/0/8]port link-type access
Info: This operation may take a few seconds. Please wait for a moment...done.
[SWA-GigabitEthernet0/0/9]port link-type access
Info: This operation may take a few seconds. Please wait for a moment...done.
[SWA-GigabitEthernet0/0/10]port link-type access
Info: This operation may take a few seconds. Please wait for a moment...done.
[SWA-port-group]port default vlan 20   //将GE0/0/6～GE0/0/10接口划分到VLAN20
[SWA-GigabitEthernet0/0/6]port default vlan 20
[SWA-GigabitEthernet0/0/7]port default vlan 20
[SWA-GigabitEthernet0/0/8]port default vlan 20
[SWA-GigabitEthernet0/0/9]port default vlan 20
[SWA-GigabitEthernet0/0/10]port default vlan 20
[SWA-port-group]quit
```

5. 将交换机 SWA 与路由器 RTA 的互连接口 GE0/0/24 配置为 Trunk（中继）模式，允许所有 VLAN 数据流量进入路由器 RTA。

```
[SWA]interface GigabitEthernet 0/0/24   //进入交换机的GE0/0/24接口的端口配置状态
[SWA-GigabitEthernet0/0/24]port link-type trunk//配置GE0/0/24接口的端口类型为Trunk
[SWA-GigabitEthernet0/0/24]port trunk allow-pass vlan 10 20   //配置GE0/0/24接口的Trunk
```

模式，允许VLAN10、VLAN20数据通过
　　　[SWA-GigabitEthernet0/0/24]quit　//退出端口配置模式，返回全局配置模式

🔥 小贴士

　　华为交换机默认不支持 VLAN 1 之外的其他 VLAN 通过，所以要允许 Trunk 端口通过 VLAN10、VLAN20 的流量。

6. 保存配置，保证交换机重启之后不会丢失配置。

```
[SWA]quit  //退出全局配置模式，返回特权用户模式
<SWA>save  //保存当前所做配置
The current configuration will be written to the device.
Are you sure to continue?[Y/N]y  //此处输入yes，确认后配置被保存
#提示：修改配置对于交换机来说是一个危险操作，系统要求进行确认
 Info: Please input the file name ( *.cfg, *.zip ) [vrpcfg.zip]:  //系统要求在冒号后面输入配置文件的名字，[]里给出了一个系统默认的文件名
 Apr  2 2000 00:46:14 SWA %%01CFM/4/SAVE(s)[0]:The user chose Y when deciding whether to save the configuration to the device.  //按Enter键，选择采用默认的文件名
#提示：沿用默认的文件名保存当前配置，将覆盖之前已经存在的相同名称的配置文件
flash:/vrpcfg.zip exists, overwrite?[Y/N]:yes  //系统提示是否覆盖默认的配置文件，输入yes确认覆盖
 Apr  2 2000 00:46:36 SWA %%01CFM/4/OVERWRITE_FILE(s)[1]:When deciding whether to overwrite the configuration file vrpcfg.zip, the user chose Y.
 Now saving the current configuration to the slot 0...
 Save the configuration successfully.    //系统提示保存成功
```

🔥 小贴士

　　在交换机上输入命令后立即生效，并将命令保存在交换机当前的运行内存中，如果遇到故障或运行错误，则在交换机重启后这些命令就无效了。为了使命令或配置在交换机重新加电启动后还能生效，则需要使用 save 命令将这些命令或配置保存到固定存储器中。

7. 在路由器 RTA 上配置子接口，并封装对应的 VLAN 流量。

```
<Huawei>system-view    //进入路由器系统视图
Enter system view, return user view with Ctrl+Z.
#提示：同时按<Ctrl>+<Z>键，可返回用户视图
[Huawei]sysname RTA   //当前路由器默认名为Huawei，修改其设备名称为RTA
[RTA]interface GigabitEthernet 0/0/1.10  //进入GE0/0/1.10子接口
#提示：接口名称后加.10，表示此子接口名称为x.10，以下为其他子接口
[RTA-GigabitEthernet0/0/1.10]dot1q termination vid 10 //封装dot1q协议,该子接口对应为VLAN10承载流量
#提示：dot1q表示按照802.1q协议运行，把该子接口分配给VLAN10承载流量
[RTA-GigabitEthernet0/0/1.10]ip address 192.168.10.254 24    //设置子接口IP地址和子网掩码，24表示子网掩码255.255.255.0
[RTA-GigabitEthernet0/0/1.20]arp broadcast enable   //开启子接口的ARP广播
#提示：必须开启广播，此子接口才能作为该VLAN的网关工作
```

```
    [RTA-GigabitEthernet0/0/1.10]quit          //退出子接口GE0/0/1.10配置视图
    [RTA]interface GigabitEthernet 0/0/1.20    //进入子接口GE0/0/1.20配置视图
    [RT1-GigabitEthernet0/0/1.20]dot1q termination vid 20    //封装dot1q协议,该子接口对应
VLAN20
    #提示:dot1q表示按照802.1q协议运行,把该子接口分配给VLAN20承载流量
    [RT1-GigabitEthernet0/0/1.20]ip address 192.168.20.254 255.255.255.0
//设置子接口IP地址和子网掩码
    [RTA-GigabitEthernet0/0/1.20]arp broadcast enable    //开启子接口的ARP广播,允许其转发
VLAN广播
    #提示:必须开启广播,此子接口才能作为该VLAN的网关工作
    [RT1-GigabitEthernet0/0/1.20]quit    //退出子接口配置视图
    [RTA]quit    //退出系统视图
```

8. 输入 save 命令，保存所做配置，使在路由器重启后配置不会丢失。

```
    <RTA>save    //保存所做配置
    #提示:save命令在用户视图中使用,保存命令到路由配置文件中,保证设备重启后仍然有效
    Warning: The current configuration will be written to the device.
    Are you sure to continue?[Y/N]:y    //输入yes 确认保存
    #提示:修改配置对于路由器来说是一个危险操作,系统要求进行确认
    It will take several minutes to save configuration file, please wait............
    Configuration file had been saved successfully
    Note: The configuration file will take effect after being activated
//保存成功
    <RTA>    //系统进入用户视图
```

9. 配置完成后使用 ping 命令测试 PC1 与 PC2 的连通性，如图 3-4-4 所示。

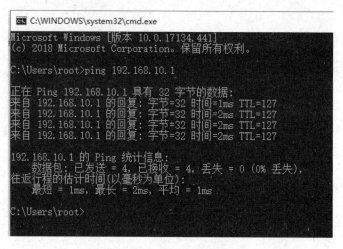

图 3-4-4　连通性测试效果

 任务总结与思考

本任务重点讲述了使用路由器实现不同 VLAN 间的通信，在实现过程中，重点理解子接口的概念。

思考以下两个问题。

1. 在以太网 GE0/0/1 接口中是否需要配置 IP 地址？此种情况下的物理接口在配置封装

之后是仅作为一个二层的链路通道，还是作为一个正常功能的三层接口？

2. 测试时是否可以使用 VLAN1 的成员进行测试？单臂路由是否可以使用 Trunk 接口与主 VLAN 成员连通？

 知识补给

单臂路由模式，即在路由器上设置多个逻辑子接口，每个子接口都对应一个 VLAN。每个子接口上的数据在物理链路上传递时都要标记封装。

对于路由器的接口，在支持子接口的同时，还必须支持 Trunk 功能。使用单臂路由模式配置 VLAN 间路由时，路由器的物理接口必须与相邻交换机的 Trunk 链路相连。在路由器上，每个子接口都会分配专属于其子网 VLAN 的唯一 IP 地址，同时也便于为该 VLAN 标记帧。这样，路由器就可以在流量通过 Trunk 链路且返回交换机时区分不同子接口的流量。

以下举例说明单臂路由的命令行配置方法。

```
[RT1]interface GigabitEthernet 0/0/1.20    //进入子接口GE0/0/1.20
[RT1-GigabitEthernet0/0/1.20]dot1q termination vid 20    //封装dot1q协议,该子接口对应承载VLAN20流量
[RT1-GigabitEthernet0/0/1.20]ip address 192.168.20.254 255.255.255.0    //设置子接口IP地址和子网掩码
[RTA-GigabitEthernet0/0/1.20]arp broadcast enable    //开启子接口的ARP广播
[RT1-GigabitEthernet0/0/1.20]quit
[RTA]quit
```

3.5 升级路由器

➢ 任务情景

宇信公司的企业网络中采用了一批 HUAWEIAR2220E 路由器。经过一段时间的运行，小李发现个别设备的转发效率偏低，数据吞吐量没有达到项目设计的预期目标。为了优化路由器性能，他决定对路由器操作系统进行升级，具体方法是使用 PC 作为 TFTP 服务器来提供升级服务。

➢ 任务分析

- ➢ 了解路由器操作系统版本；
- ➢ 学会从官网下载所需软件；
- ➢ 学会设置并连接 TFTP 服务器。

➢ 实施准备

1. HUAWEIAR2220E 路由器 1 台；

2．PC 1 台；

3．交叉型网线 1 根；

4．Console 通信线缆 1 根；

5．TFTP 软件 1 套；

6．新版操作系统 1 套。

【提示：TFTP 即简单文件传输协议，互联网上有多种软件支持这个协议。】

小贴士

路由器具体型号可通过 display version 命令查看，并可从华为公司官网下载该路由器的操作系统。本次实验所用路由器型号为 HUAWEIAR2220E。

➢ 实施步骤

1．按任务要求连接网络设备，升级路由器拓扑图如图 3-5-1 所示。

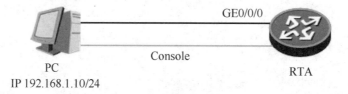

图 3-5-1　升级路由器拓扑图

2．按任务情景需要规划路由器 RTA、PC 的 IP 地址及 Mask，设备地址分配表如表 3-5-1 所示。

表 3-5-1　设备地址分配表

设备名称	接口	IP 地址	Mask	网关
RTA	GE0/0/0	192.168.1.1	255.255.255.0	无
PC1	Ethernet	192.168.1.10	255.255.255.0	无

3．登录路由器，查看当前版本。

```
< Huawei>display version    //查看路由器操作系统版本
Huawei Versatile Routing Platform Software
 VRP (R) software, Version 5.160 (AR2200 V200R007C00SPC900) //显示当前操作系统版本为AR2200平台的V200R007C00SPC900
 Copyright (C) 2011-2016 HUAWEI TECH CO., LTD   //当前版本的发行时间
 Huawei AR2220E Router uptime is 0 week, 6 days, 23 hours, 41 minutes  //路由器持续更新时间

 BKP  0 version information:             //背板信息
 1. PCB      Version     : AR01BAK2C VER.B   //PCB版本信息
 2. If Supporting PoE : No            //是否支持PoE功能
 3. Board    Type      : AR2220E       //主板型号为AR2220E
 4. MPU Slot Quantity : 1            //主板槽位数量为1
 5. LPU Slot Quantity : 6            //接口板卡槽位数量为6
```

```
     MPU 0(Master) : uptime is 0 week, 6 days, 23 hours, 40 minutes
//主板上电时间已持续6天23小时40分钟
     SDRAM Memory Size    : 1024    M bytes        //SDRAM内存的容量
     Flash 0 Memory Size  : 512     M bytes        //Flash闪存的容量
     Flash 1 Memory Size  : 16      M bytes
     NVRAM Memory Size    : 512     K bytes        //非易失性存储容量
     MPU version information :                     //主板版本信息
     1. PCB       Version      : AR-SRU2220E VER.A
     2. MAB       Version      : 0
     3. Board     Type         : AR2220E
     4. CPLD0     Version      : 100
     5. BootROM   Version      : 312

     LPU 5 : uptime is 0 week, 6 days, 23 hours, 39 minutes
       //第5个槽位上的接口板上电时间
     SDRAM Memory Size    : 256     M bytes        //SDRAM内存容量
     Flash 0 Memory Size  : 16      M bytes        //Flash的容量
     LPU version information :                     //LPU版本信息
     1. PCB       Version      : AR01WDAS8A VER.C
     2. MAB       Version      : 0
     3. Board     Type         : 8AS
     4. BootROM   Version      : 502
```

4. 在 PC 资源管理器中找到已下载的 TFTP 软件，TFTP 软件解压缩前的文件名及压缩包如图 3-5-2 所示。右击该压缩包，在弹出的快捷菜单中选择 "解压到 Tftpd32_cn\(E)"，将生成一个名为 Tftpd32_cn 的文件夹，如图 3-5-3 所示。

图 3-5-2　解压缩运行前的 TFTP 软件　　图 3-5-3　解压缩运行后的 TFTP 文件夹

5. 打开文件夹，选择 Tftpd32 汉化版，在如图 3-5-4 所示的界面中双击运行可执行文件 tftpd32.exe，打开 TFTP 应用窗口，如图 3-5-5 所示。

图 3-5-4　TFTP 可执行文件位置

6. 在 TFTP 应用窗口上方的"服务器地址"下拉列表里选定当前 PC 的 IP 地址，在"当前目录"下拉列表里指定文件存放位置，当前 PC 即可作为 TFTP 服务器，如图 3-5-6 所示。

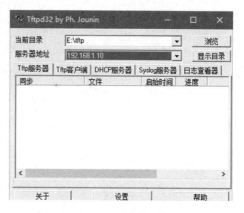

图 3-5-5　TFTP 应用窗口　　　　　　图 3-5-6　TFTP 服务器设置

7. 访问华为公司官网，依次单击"技术支持"→"路由器"→"接入路由器"→"AR2200E"→"软件"，下载路由器的新版操作系统，并将其保存到 TFTP 服务器工作目录下，如图 3-5-7 所示。

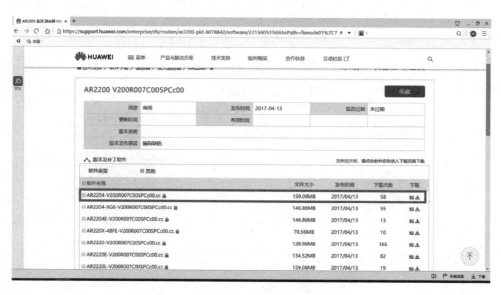

图 3-5-7　新版操作系统源文件信息

8. 登录路由器，从 TFTP 服务器中获取最新版系统，TFTP 传输文件过程如图 3-5-8 所示，路由器传输界面如图 3-5-9 所示。

```
< Huawei > tftp 192.168.1.10 get AR2220E-V200R007C00SPCc00.cc
//从TFTP服务器上获取新版操作系统
#提示：确认服务器的IP地址、软件的名称及服务器与AR路由器之间连通，文件传输才能成功
Info: Transfer file in binary mode.
Downloading the file from the remote TFTP server. Please wait...
#提示：文件正在下载，请等待
99%
141052928 bytes received in 399 seconds.
TFTP: Downloading the file successfully.
```

#提示：141052928字节的文件在399秒内完成传输，文件下载已成功

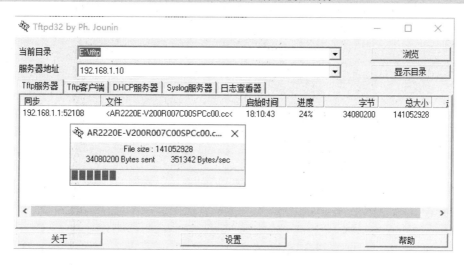

图 3-5-8　TFTP 传输文件过程

```
<Huawei>tftp 192.168.1.10 get AR2220E-V200R007C00SPCc00.cc
Info: Transfer file in binary mode.
Downloading the file from the remote TFTP server. Please wait...
 26%
```

图 3-5-9　路由器传输界面

9. 在路由器上检查文件是否上传成功，文件大小是否和服务器上的文件完全相同。

```
<Huawei>dir flash:     //查看路由器文件系统信息
Directory of flash:/

  Idx  Attr     Size(Byte)  Date       Time(LMT)    FileName
    0  -rw-    126,538,240  Oct 12 2017 14:54:20   AR2220E-V200R007C00SPC900.cc
    1  -rw-         25,603  Dec 16 2018 21:12:06   mon_file.txt
    2  -rw-            783  Dec 09 2018 20:57:40   default_local.cer
    3  -rw-            863  Dec 12 2018 19:23:06   vrpcfg.zip
    4  -rw-              0  Mar 05 2018 17:35:32   brdxpon_snmp_cfg.efs
    5  drw-              -  Mar 05 2018 17:36:02   update
    6  drw-              -  Mar 05 2018 17:36:26   shelldir
    7  drw-              -  Mar 05 2018 17:36:38   localuser
    8  drw-              -  Mar 08 2018 04:46:22   dhcp
    9  -rw-      5,899,648  Mar 09 2018 23:01:12   AR2220E-V200R007SPH013.pat
   10  -rw-            494  Dec 12 2018 19:23:08   private-data.txt
   11  -rw-            902  Dec 12 2018 16:18:46   mon_lpu_file.txt
   12  -rw-          1,024  Dec 16 2018 21:09:02   Boot_LogFile
   13  -rw-    141,052,928  Dec 19 2018 11:11:30   AR2220E-V200R007C00SPCc00.cc

510,484 KB total available (243,276 KB free)
<Huawei>
```

#提示：根据路由器反馈信息可知，从TFTP服务器下载的新版操作系统的文件名为 AR2220E-V200R007C00SPCc00.cc，字节数为141052928，与源文件相同，可以正常使用

10. 设置 AR 路由器的默认启动系统软件为新上传的系统软件。

```
<Huawei>startup system-software flash:/AR2220E-V200R007C00SPCc00.cc   //设置系统从
flash目录中的AR2220E-V200R007C00SPCc00.cc文件启动
Info: This operation will take several minutes, please wait.......
```

#提示：系统提示此操作需要几分钟，请等待

```
Info: Succeeded in setting the file for main booting system
#提示：系统提示已成功地引导该系统
```

> **小贴士**
>
> 在系统中进行输入操作时，若命令行或文件名较长，则可只输入前几个字符，再按<TAB>键将信息补全。

11. 重新启动路由器系统。

```
<Huawei>reboot  //重启路由器系统
Info: The system is comparing the configuration, please wait.
Warning:All the configuration will be saved to the next startup configuration. Continue?
[y/n]:y
  It will take several minutes to save configuration file, please wait..........
  Configuration file had been saved successfully
  Note: The configuration file will take effect after being activated
System will reboot! Continue? [y/n]:y
#提示：路由器系统将要重启，请输入yes确认
Info: system is rebooting, please wait...
Dec 19 2018 11:20:17+00:00 Huawei ENTITYTRAP/4/BOARDINVALID:OID
1.3.6.1.4.1.2011.5.25.219.2.2.5 Board is invalid for some reason.(Index=1310729,
EntityPhysicalIndex=1310729, PhysicalName="Board 5", EntityTrapEntType= 2,
EntityTrapFaultID=132627, EntityTrapReasonDescr="Board Registration Failed")
<Huawei>
    Dec 19 2018 11:20:18+00:00 Huawei ENTMIB/4/TRAP:OID 1.3.6.1.2.1.47.2.0.1 Entity MIB
changes.
    … // 此处省略一些非必要系统信息
<Huawei>
<Huawei>[223842.145136] Restarting system.


************************************************************
***                   HUAWEI AR.                        ***
***BOOT version:312 (Build time: May 10 2016 - 15:58:51) ***
************************************************************

..................
Check File in patch flash:/AR2220E-V200R007SPH013.pat
#提示：以上显示信息为路由器系统硬件版本及当前待升级操作系统相关信息
Now boot from flash:/AR2220E-V200R007C00SPCc00.cc, please wait...
#提示：即将加载从flash:/中加载新版操作系统，请等待……
Check file :Boot .
Current Bootrom Version   : 0x312
Bootrom in packet Version : 0x311

Update Bootrom ... ........OK
Check file :CPLD .OK


System will reboot
...
************************************************************
***                   HUAWEI AR.                        ***
***BOOT version:311 (Build time: Apr 29 2016 - 18:18:17) ***
```

```
************************************************************
Press ctrl+T for memory test  .... skip
Flash  : 16 MiB
SD card: 0 Device(s) found
USB    : 3 USB Device(s) found
       : 1 Storage Device(s) found
Net    : octeth0, octeth1, octeth2, octeth3, octeth4, octeth5, octeth6, octeth7

Press Ctrl+A for Bootrom menu ... 0
Now boot from flash:/AR2220E-V200R007C00SPCc00.cc, please wait...
Check file :Boot .OK
Check file :CPLD .OK
Check file :cn6xxx_rootfs ....
Current file Version   : 0x303331
file in packet Version : 0x353331
update file ....................................OK
Check file :cn6xxx_kernel ..
Current file Version   : 0x373939
file in packet Version : 0x303930
update file ..............................OK
Check file :bootloaderfs ..OK
Starting system
start from app_squashfs.
... OK
Press Ctrl+B to break auto startup ... 0
*********EVM Init(Version:D20141206B02)*********
EVM execute start monitor(0)......OK.

Create monitor process

CPLD userspace init ok

INFO:Get pri len, the result is 0x810401ab
Link voice so to ag.so
Create cap process
Create vrp process

CPLD userspace init ok

Initializing SMI Bus:OK

Initializing I2C Bus:OK
Backup Bootrom : done

DOPRA initialize..................OK
IAS initialize....................OK

VRP_SockTm_Init create............OK
VRP initialize
Create tasks......................OK
Initialize tasks..................OK
Recovering configuration..........OK
```

```
System will reboot
#提示：系统即将重启
···//以下省略部分系统硬件自检信息

CPLD userspace init ok
#提示：系统初始化结束
Initializing SMI Bus:OK

Initializing I2C Bus:OK
Backup Bootrom : done

DOPRA initialize..................OK
IAS initialize....................OK

VRP_SockTm_Init create...........OK
VRP initialize
#提示：新版VRP软件初始化完成
Create tasks.....................OK
Initialize tasks.................OK
Recovering configuration.........OK
#提示：成功恢复系统升级前配置
<Huawei>
```

12. 检查操作系统版本是否为新上传版本 5.160。

```
<Huawei>display version
Huawei Versatile Routing Platform Software
VRP (R) software, Version 5.160 (AR2200 V200R007C00SPCc00)
#提示：此处显示操作系统版本已为更新版本
Copyright (C) 2011-2017 HUAWEI TECH CO., LTD
···//省略部分系统提示信息
<Huawei>
```

任务总结与思考

本任务重点讲述利用 TFTP 服务器对路由器操作系统进行升级的完整操作过程。

思考以下两个问题。

1．在此升级过程中，PC 与路由器之间为什么要通过 2 根线缆连接？它们分别起到什么任用？

2．本任务多次重启路由器，请描述路由器启动的步骤。

任务拓展

下面我们通过另一种方法，在保持任务拓扑不变的情况下，在 PC 上安装 FTP 软件后将其升级为 FTP 服务器，路由器作为 FTP 客户端，不需要进行任何配置即可实现升级。

1．在网上搜索并下载 FTP 软件 3CDaemon，并保存在 PC 桌面，然后将软件解压缩到指定位置，FTP 软件压缩包及解压缩文件夹如图 3-5-10 所示。

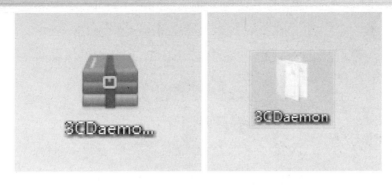

图 3-5-10　FTP 软件压缩包及解压缩文件夹

2．打开解压缩文件夹，找到可执行文件，如图 3-5-11 所示。

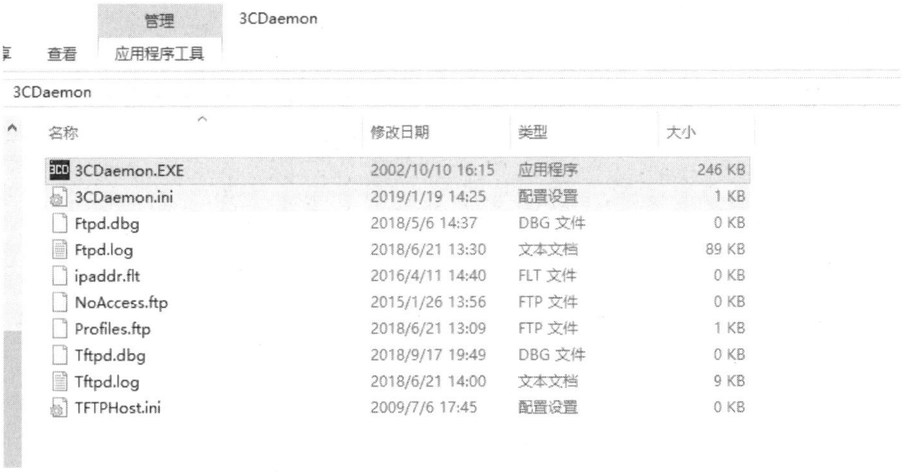

图 3-5-11　选择 FTP 可执行文件

3．双击运行可执行文件，打开 FTP 应用窗口，如图 3-5-12 所示。

图 3-5-12　FTP 应用窗口

4. 由于本软件是一个多功能软件，因此需要首先选择"文件"→"配置选定的服务"选项，如图 3-5-13 所示。在弹出的对话框中逐级配置 FTP 用户的工作文件夹，如图 3-5-14 所示。

图 3-5-13　选择配置的服务

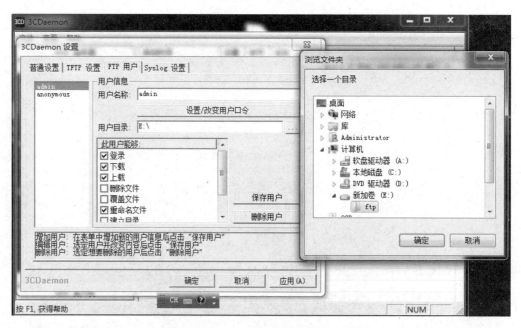

图 3-5-14　配置 FTP 工作文件夹位置

5. 将准备启用的新的路由器系统文件保存至指定的工作目录，如图 3-5-15 所示。

图 3-5-15　将新的系统文件保存到 FTP 工作目录

6. 输入命令，尝试连接 FTP 服务器。

```
<Huawei>ftp 192.168.1.10  //路由器尝试连接FTP服务器
Trying 192.168.1.10 ...
Press CTRL+K to abort
Connected to 192.168.1.10 .
```

```
220 FTP Server ready.
User(192.168.1.10 :(none)):admin      //输入用户名,按Enter键
331 Password required for AR.
Enter password:
#提示:如果已设置FTP用户密码,则需要验证;若无密码,则直接按Enter键继续
230 User 1 logged in.
#提示:登录成功
```

7. 指定 FTP 传输格式为二进制。

```
[Huawei-ftp] binary    //设置为二进制传输模式
200 Type set to I.
#提示:网络设备的系统文件通常都是以二进制形式存储的,所以下载前需要指定传输格式为二进制形式
```

8. 连接成功后,从 FTP 服务器将系统文件下载到路由器。

```
[Huawei-ftp] get AR2220E-V200R007C00SPCc00.cc    //下载系统文件
#提示:get命令的作用是将系统文件从服务器下载到客户机。如果需要把文件从路由器上传到服务器,则需用put
<文件名>命令
200 Port command successful.
#提示:系统传输成功
150 Opening data connection for AR2220E-V200R007C00SPCc00.cc.
226 File sent ok
FTP: 69146624 byte(s) received in 265.717 second(s) 254.13Kbyte(s)/sec.
#提示:FTP提示传输了69146624字节,与工作目录下的文件大小一致,表示传输成功
```

9. 在路由器上指定以新的操作系统启动。

小贴士

FTP 服务是建立在 TCP 的 21 接口上的,服务器和客户端是有固定连接的,文件传输的可靠性比较高,因此被广泛采用。TFTP 服务是建立在 UDP 的 69 接口上的,没有固定连接,优点是比较灵活,多用于网络设备系统升级。

3.6 访问外部网络

访问外部网络

➤ 任务情景

宇信公司的新办公网出口路由器为 RTC,原驻地网络路由器为 RTA,两者之间跨越一个运营商的路由器 RTB,目前都未配置任何动态路由协议,导致原驻地网络数据与新办公地数据无法共享。由于涉及的路由比较单一,所以小李决定尝试通过在路由器上配置静态路由来实现网络互通。

➤ 任务分析

- ➤ 学会配置路由器的静态路由;
- ➤ 学会测试静态路由的连通性;
- ➤ 掌握通过配置静态路由访问外部网络。

➢ 实施准备

1. HUAWEI AR2220E 路由器 3 台；
2. PC 2 台；
3. 交叉型网线 4 根；
4. Console 通信线缆 1 根。

➢ 实施步骤

1. 按任务要求连接网络设备，静态路由配置拓扑图如图 3-6-1 所示。

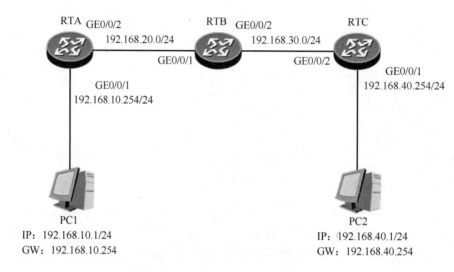

图 3-6-1 静态路由配置拓扑图

2. 按任务要求规划 IP 地址、子网掩码（Mask）和网关，地址分配如表 3-6-1 所示。PC1 和 PC2 的地址配置分别如图 3-6-2 和图 3-6-3 所示。

表 3-6-1 设备地址分配表

设备名称	接口	IP 地址	Mask	网关
RTA	GE0/0/1	192.168.10.254	255.255.255.0	无
RTA	GE0/0/2	192.168.20.1	255.255.255.0	无
RTB	GE0/0/1	192.168.20.2	255.255.255.0	无
RTB	GE0/0/2	192.168.30.1	255.255.255.0	无
RTC	GE0/0/2	192.168.30.2	255.255.255.0	无
RTC	GE0/0/1	192.168.40.254	255.255.255.0	无
PC1	Ethernet	192.168.10.1	255.255.255.0	192.168.10.254
PC2	Ethernet	192.168.40.1	255.255.255.0	192.168.40.254

三层业务互访　项目 3

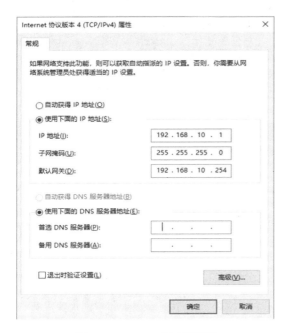

图 3-6-2　PC1 地址配置　　　　　　图 3-6-3　PC2 地址配置

3．配置路由器 RTA 的名称及相关接口 IP 地址。

```
<Huawei>system-view  //进入系统视图
Enter system view, return user view with Ctrl+Z.
#提示：返回用户视图，请同时按<Ctrl>+<Z>键
[Huawei]sysname RTA     //路由器启动后默认设备名称为Huawei，此处修改为RTA
[RTA]interface GigabitEthernet 0/0/1     //进入GE0/0/1接口配置视图
[RTA-GigabitEthernet0/0/1]ip address 192.168.10.254 24  //配置GE0/0/1接口IP地址为
192.168.10.254，并设置掩码为24位
#提示：24位掩码为255.255.255.0
Dec 12 2018 16:32:24+00:00 RTA %%01IFNET/4/LINK_STATE(l)[0]:The line protocol IP on
the interface GigabitEthernet0/0/1 has entered the UP state.
#提示：此处系统提示GE0/0/1接口的协议状态为UP，说明接口配置成功
[RTA-GigabitEthernet0/0/1]quit   //退出GE0/0/1接口配置视图
[RTA]interface GigabitEthernet 0/0/2     //进入GE0/0/2接口配置视图
[RTA-GigabitEthernet0/0/2]ip address 192.168.20.1 24   //配置GE0/0/2接口IP地址为
192.168.20.1，并设置掩码为255.255.255.0
Dec 12 2018 16:32:58+00:00 RTA %%01IFNET/4/LINK_STATE(l)[1]:The line protocol IP on
the interface GigabitEthernet0/0/2 has entered the UP state.
#提示：此处系统提示GE0/0/2接口的协议状态为UP，说明接口配置成功
[RTA-GigabitEthernet0/0/2]quit   //退出GE0/0/2接口配置视图
[RTA]  //返回系统视图
```

4．配置路由器 RTB 的名称及互连接口 IP 地址。

```
<Huawei>system-view  //进入系统视图
Enter system view, return user view with Ctrl+Z.
#提示：返回用户视图，请同时按<Ctrl>+<Z>键
[Huawei]sysname RTB     //把默认设备名称Huawei修改为RTB
[RTB]interface GigabitEthernet 0/0/1     //进入GE0/0/1接口配置视图
[RTB-GigabitEthernet0/0/1]ip address 192.168.20.2 24   //配置GE0/0/1接口的IP地址为
192.168.20.2，并设置掩码为255.255.255.0
Dec 12 2018 16:45:39+00:00 RTB %%01IFNET/4/LINK_STATE(l)[0]:The line protocol IP on
```

```
the interface GigabitEthernet0/0/1 has entered the UP state.
        #提示：此处系统提示GE0/0/1接口的协议状态为UP，说明接口配置成功
        [RTB-GigabitEthernet0/0/1]quit    //退出GE0/0/1接口配置视图
        [RTB]interface GigabitEthernet 0/0/2    //进入GE0/0/2接口配置视图
        [RTB-GigabitEthernet0/0/2]ip address 192.168.30.1 24    //配置GE0/0/2接口的IP地址为
192.168.30.1，并设置掩码为255.255.255.0
        Dec 12 2018 16:45:53+00:00 RTB %%01IFNET/4/LINK_STATE(l)[1]:The line protocol IP on
the interface GigabitEthernet0/0/2 has entered the UP state.
        #提示：此处系统提示GE0/0/2接口的协议状态为UP，说明接口配置成功
        [RTB-GigabitEthernet0/0/2]quit
        [RTB]
```

5. 配置路由器 RTC 的名称及互连接口 IP 地址。

```
        <Huawei>system-view    //进入系统视图
        Enter system view, return user view with Ctrl+Z.
        #提示：返回用户视图，请同时按<Ctrl>+<Z>键
        [Huawei]sysname RTC    //把默认设备名称Huawei修改为RTC
        [RTC]interface GigabitEthernet 0/0/1    //进入GE0/0/1接口配置视图
        [RTC-GigabitEthernet0/0/1]ip address 192.168.40.254 24    //配置此接口IP地址为
192.168.40.254，并设置掩码为255.255.255.0
        Dec 12 2018 16:54:43+00:00 RTC %%01IFNET/4/LINK_STATE(l)[0]:The line protocol IP on
the interface GigabitEthernet0/0/2 has entered the UP state.
        #提示：此处系统提示GE0/0/1接口的协议状态为UP，说明接口配置成功
        [RTC-GigabitEthernet0/0/1]quit    //退出GE0/0/1接口配置视图
        [RTC]interface GigabitEthernet 0/0/2    //进入GE0/0/2接口配置视图
        [RTC-GigabitEthernet0/0/2]ip address 192.168.30.2 24    //配置接口IP地址为192.168.30.2，
并设置掩码为255.255.255.0
        Dec 12 2018 16:54:43+00:00 RTC %%01IFNET/4/LINK_STATE(l)[0]:The line protocol IP on
the interface GigabitEthernet0/0/2 has entered the UP state.
        #提示：此处系统提示GE0/0/2接口的协议状态为UP，说明接口配置成功
        [RTC-GigabitEthernet0/0/2]quit    //退出GE0/0/2接口配置视图
        [RTC]    //返回系统视图
```

6. 查看路由器 RTA 的接口状态和路由表时，能够看到当前路由器的相关接口物理状态和协议状态均为 UP，路由表中产生两条直连路由。

```
        [RTA]display ip interface brief    //查看接口的简要信息
        ...    //省略部分屏幕显示信息
        Interface                    IP Address/Mask       Physical    Protocol
        GigabitEthernet0/0/0         192.168.1.1/24        down        down
        GigabitEthernet0/0/1         192.168.10.254/24     up          up
        GigabitEthernet0/0/2         192.168.20.1/24       up          up
        GigabitEthernet0/0/3         unassigned            up          down
        ...    //省略其他无关接口信息
        #提示：此处系统提示GE0/0/1和GE0/0/2接口的IP地址分别为192.168.10.254和192.168.20.1，与表3.6.1中规划
一致。physical（物理）状态和protocol（协议）状态均为UP，说明接口配置成功
        [RTA]display ip routing-table    //查看RTA的路由表
        Route Flags: R - relay, D - download to fib
        ------------------------------------------------------------------
        Routing Tables: Public
                Destinations : 10      Routes : 10

        Destination/Mask    Proto    Pre    Cost    Flags NextHop         Interface
```

```
  127.0.0.0/8      Direct  0    0              D   127.0.0.1        InLoopback0
  127.0.0.1/32     Direct  0    0              D   127.0.0.1        InLoopback0
  127.255.255.255/32  Direct  0   0            D   127.0.0.1        InLoopback0
  192.168.10.0/24  Direct  0    0              D   192.168.10.254   GE0/0/1
  192.168.10.254/32 Direct 0    0              D   127.0.0.1        GE0/0/1
  192.168.10.255/32 Direct 0    0              D   127.0.0.1        GE0/0/1
  192.168.20.0/24  Direct  0    0              D   192.168.20.1     GE0/0/2
  192.168.20.1/32  Direct  0    0              D   127.0.0.1        GE0/0/2
  192.168.20.255/32 Direct 0    0              D   127.0.0.1        GE0/0/2
  255.255.255.255/32 Direct 0   0              D   127.0.0.1        InLoopback0
[RTA]
```

#提示：注意此时路由器RTA的路由表中出现了192.168.10.0/24和192.168.20.0/24的路由，协议（Proto）类型均为直连（Direct）类型，下一跳（FlagsNextHop）地址分别为192.168.10.254和192.168.20.1，即路由器的GE0/0/1和GE0/0/2接口，说明RTA可以正常转发去往这两个网络的数据包

#此时可以验证路由器RTA与其他网络设备的连通性。从以下输出结果可以看出，路由器RTA与路由器RTB及PC1都已连通，但不能访问路由器RTC及PC2

```
[RTA]ping 192.168.10.1
  PING 192.168.10.1: 56  data bytes, press CTRL_C to break
    Reply from 192.168.10.1: bytes=56 Sequence=1 ttl=128 time=1 ms
    Reply from 192.168.10.1: bytes=56 Sequence=2 ttl=128 time=1 ms
    Reply from 192.168.10.1: bytes=56 Sequence=3 ttl=128 time=1 ms
    Reply from 192.168.10.1: bytes=56 Sequence=4 ttl=128 time=1 ms
    Reply from 192.168.10.1: bytes=56 Sequence=5 ttl=128 time=1 ms

  --- 192.168.10.1 ping statistics ---
    5 packet(s) transmitted
    5 packet(s) received
    0.00% packet loss
    round-trip min/avg/max = 1/1/1 ms
```

#提示：路由器RTA通过ping命令向IP为192.168.10.1的PC1发送5个ICMP包，收到了5个答复（Reply），每个包的字节数（bytes）都为56，序列号（Sequence）分别为1~5，生存时间（ttl）为128，耗费时间（time）为1ms，丢包率为0%，说明二者之间的网络已经互通，状态良好

```
[RTA]ping 192.168.20.2
  PING 192.168.20.2: 56  data bytes, press CTRL_C to break
    Reply from 192.168.20.2: bytes=56 Sequence=1 ttl=255 time=1 ms
    Reply from 192.168.20.2: bytes=56 Sequence=2 ttl=255 time=1 ms
    Reply from 192.168.20.2: bytes=56 Sequence=3 ttl=255 time=1 ms
    Reply from 192.168.20.2: bytes=56 Sequence=4 ttl=255 time=1 ms
    Reply from 192.168.20.2: bytes=56 Sequence=5 ttl=255 time=1 ms

  --- 192.168.20.2 ping statistics ---
    5 packet(s) transmitted
    5 packet(s) received
    0.00% packet loss
    round-trip min/avg/max = 1/1/1 ms
```

#提示：路由器RTA通过ping命令向IP为192.168.20.2的RTB发送5个ICMP包，收到了5个答复（Reply），每个包的字节数（bytes）都为56，序列号（Sequence）分别为1~5，生存时间（ttl）为255，耗费时间（time）为1ms，丢包率为0%，说明二者之间的网络互通，状态良好

```
[RTA]ping 192.168.30.2
  PING 192.168.30.2: 56  data bytes, press CTRL_C to break
    Request time out
    Request time out
```

```
        Request time out
        Request time out
        Request time out

    --- 192.168.30.2 ping statistics ---
        5 packet(s) transmitted
        0 packet(s) received
        100.00% packet loss
```
#提示：路由器RTA通过ping命令向IP为192.168.30.2的路由器RTC发送5个ICMP包，未在规定时间内收到答复（Reply），系统反馈5个请求包均超时（Request time out），丢包率为100%，说明二者之间的网络未连通
```
    [RTA]ping 192.168.40.1
    PING 192.168.40.1: 56  data bytes, press CTRL_C to break
        Request time out
        Request time out
        Request time out
        Request time out
        Request time out

    --- 192.168.40.1 ping statistics ---
        5 packet(s) transmitted
        0 packet(s) received
        100.00% packet loss
    [RTA]
```
#提示：路由器RTA通过ping命令向IP为192.168.40.1的PC2发送5个ICMP包，未在规定时间内收到答复（Reply），系统反馈请求超时（Request time out），丢包率为100%，说明二者之间的网络未连通

7. 查看路由器 RTB 的接口和路由表时，能够看到当前路由器的相关接口物理状态和协议状态均为 UP，路由表中产生两条直连路由。

```
    [RTB]display ip interface brief   //查看路由器RTB的接口简要信息
    ...  //省略部分与本任务无关的屏幕提示信息

Interface                    IP Address/Mask      Physical     Protocol
GigabitEthernet0/0/0         192.168.1.1/24       down         down
GigabitEthernet0/0/1         192.168.20.2/24      up           up
GigabitEthernet0/0/2         192.168.30.1/24      up           up
    ...  //省略部分与本任务无关的接口信息
```
#提示：此处系统提示GE0/0/1和GE0/0/2的接口IP地址分别为192.168.20.2和192.168.30.1，与表3.6.1中规划一致。物理（physical）状态和协议（protocol）状态均为UP，说明接口配置成功
```
    [RTB]display ip routing-table
Route Flags: R - relay, D - download to fib
------------------------------------------------------------------------
Routing Tables: Public
         Destinations : 10        Routes : 10

Destination/Mask       Proto    Pre  Cost    Flags  NextHop         Interface

    127.0.0.0/8        Direct    0    0        D    127.0.0.1       InLoopback0
    127.0.0.1/32       Direct    0    0        D    127.0.0.1       InLoopback0
    127.255.255.255/32 Direct    0    0        D    127.0.0.1       InLoopback0
    192.168.20.0/24    Direct    0    0        D    192.168.20.2    GE0/0/1
    192.168.20.2/32    Direct    0    0        D    127.0.0.1       GE0/0/1
```

```
192.168.20.255/32  Direct  0  0    D   127.0.0.1       GE0/0/1
192.168.30.0/24    Direct  0  0    D   192.168.30.1    GE0/0/2
192.168.30.1/32    Direct  0  0    D   127.0.0.1       GE0/0/2
192.168.30.255/32  Direct  0  0    D   127.0.0.1       GE0/0/2
255.255.255.255/32 Direct  0  0    D   127.0.0.1       InLoopBack0
```

[RTB]

#提示：注意此时路由器RTB的路由表中出现了192.168.20.0/24和192.168.30.0/24的路由，协议（Proto）类型均为直连(Direc)类型，下一跳 (FlagsNextHop) 地址分别为192.168.20.2和192.168.30.1，即路由器的GE0/0/1和GE0/0/2接口，说明路由器RTB可以正常转发去往这两个网络的数据包

#此时可以验证路由器RTB与其他网络设备的连通性。从如下测试输出结果可以看到，RTB与RTA及RTC都已连通，但不能访问PC1及PC2

```
[RTB]ping 192.168.20.1
  PING 192.168.20.1: 56  data bytes, press CTRL_C to break
    Reply from 192.168.20.1: bytes=56 Sequence=1 ttl=255 time=1 ms
    Reply from 192.168.20.1: bytes=56 Sequence=2 ttl=255 time=1 ms
    Reply from 192.168.20.1: bytes=56 Sequence=3 ttl=255 time=1 ms
    Reply from 192.168.20.1: bytes=56 Sequence=4 ttl=255 time=1 ms
    Reply from 192.168.20.1: bytes=56 Sequence=5 ttl=255 time=1 ms

  --- 192.168.20.1 ping statistics ---
    5 packet(s) transmitted
    5 packet(s) received
    0.00% packet loss
    round-trip min/avg/max = 1/1/1 ms
```

#提示：路由器RTB通过ping命令向IP为192.168.20.1的RTA发送5个ICMP包，收到了5个答复（Reply），每个包的字节数（bytes）都为56，序列号（Sequence）分别为1～5，生存时间（ttl）为255，耗费时间（time）为1ms，丢包率为0%，说明二者之间的网络互通，状态良好

```
[RTB]ping 192.168.30.2
  PING 192.168.30.2: 56  data bytes, press CTRL_C to break
    Request time out
    Reply from 192.168.30.2: bytes=56 Sequence=2 ttl=255 time=1 ms
    Reply from 192.168.30.2: bytes=56 Sequence=3 ttl=255 time=1 ms
    Reply from 192.168.30.2: bytes=56 Sequence=4 ttl=255 time=1 ms
    Reply from 192.168.30.2: bytes=56 Sequence=5 ttl=255 time=1 ms

  --- 192.168.30.2 ping statistics ---
    5 packet(s) transmitted
    4 packet(s) received
    20.00% packet loss
    round-trip min/avg/max = 1/1/1 ms
```

#提示：路由器RTB通过ping命令向IP为192.168.30.2的RTC发送5个ICMP包，收到了4个答复（Reply），每个包的字节数（bytes）都为56，序列号（Sequence）分别为2～5，生存时间（ttl）为255，耗费时间（time）为1ms，丢包率为20%，说明二者之间的网络已经互通

```
[RTB]ping 192.168.10.1
  PING 192.168.10.1: 56  data bytes, press CTRL_C to break
    Request time out
    Request time out
    Request time out
```

```
        Request time out
        Request time out

     --- 192.168.10.1 ping statistics ---
      5 packet(s) transmitted
      0 packet(s) received
      100.00% packet loss
```
#提示：路由器RTB通过ping命令向IP地址为192.168.10.1的PC1发送5个ICMP包，数据包的字节数(bytes)为56，却未在规定时间内收到答复(Reply)，系统反馈5个请求包超时(Request time out)，丢包率为100%，说明二者之间的网络未连通

```
    [RTB]ping 192.168.40.1
      PING 192.168.40.1: 56  data bytes, press CTRL_C to break
        Request time out
        Request time out
        Request time out
        Request time out
        Request time out

     --- 192.168.40.1 ping statistics ---
      5 packet(s) transmitted
      0 packet(s) received
      100.00% packet loss
```
#提示：路由器RTB通过ping命令向IP为192.168.40.1的PC2发送5个ICMP包，数据包的字节数(bytes)为56，却未在规定时间内收到答复(Reply)，系统反馈5个请求包超时(Request time out)，丢包率为100%，说明二者之间的网络未连通

8. 查看路由器RTC的接口和路由表时，能够看到当前路由器的相关接口物理状态和协议状态均为UP，路由表中产生两条直连路由。

```
    [RTC]display ip interface brief
    ...
    Interface                   IP Address/Mask      Physical    Protocol
    GigabitEthernet0/0/0        192.168.1.1/24       down        down
    GigabitEthernet0/0/1        192.168.40.254/24    up          up
    GigabitEthernet0/0/2        192.168.30.2/24      up          up
    ...
```
#提示：路由器RTC系统提示接口GE0/0/1和GE0/0/2的IP地址分别为192.168.40.254和192.168.30.2，与表3.6.1中规划一致。物理(physical)状态和协议(protocol)状态均为UP，说明接口配置成功

```
    [RTC]display ip routing-table
    Route Flags: R - relay, D - download to fib
    -----------------------------------------------------------------
    Routing Tables: Public
            Destinations : 10      Routes : 10

    Destination/Mask       Proto   Pre  Cost  Flags NextHop          Interface

    127.0.0.0/8            Direct  0    0     D     127.0.0.1        InLoopback0
    127.0.0.1/32           Direct  0    0     D     127.0.0.1        InLoopback0
    127.255.255.255/32     Direct  0    0     D     127.0.0.1        InLoopback0
```

```
 192.168.30.0/24       Direct  0    0         D    192.168.30.2    GE0/0/2
 192.168.30.2/32       Direct  0    0         D    127.0.0.1       GE0/0/2
 192.168.30.255/32     Direct  0    0         D    127.0.0.1       GE0/0/2
 192.168.40.0/24       Direct  0    0         D    192.168.40.254  GE0/0/1
 192.168.40.254/32     Direct  0    0         D    127.0.0.1       GE0/0/1
 192.168.40.255/32     Direct  0    0         D    127.0.0.1       GE0/0/1
 255.255.255.255/32    Direct  0    0         D    127.0.0.1       InLoopback0
[RTC]
```

#提示：注意此时路由器RTC的路由表中出现了192.168.30.0/24和192.168.40.0/24的路由，协议（Proto）类型均为直连（Direc）类型，下一跳（FlagsNextHop）地址分别为192.168.30.2和192.168.40.254，即路由器的GE0/0/2和GE0/0/1接口，说明RTC可以正常转发去往这两个网络的数据包

#此时可以验证直连设备的连通性，RTC与RTB及PC2都已连通，但不能访问RTA及PC1

```
[RTC]ping 192.168.30.1
  PING 192.168.30.1: 56  data bytes, press CTRL_C to break
    Reply from 192.168.30.1: bytes=56 Sequence=1 ttl=255 time=1 ms
    Reply from 192.168.30.1: bytes=56 Sequence=2 ttl=255 time=1 ms
    Reply from 192.168.30.1: bytes=56 Sequence=3 ttl=255 time=1 ms
    Reply from 192.168.30.1: bytes=56 Sequence=4 ttl=255 time=1 ms
    Reply from 192.168.30.1: bytes=56 Sequence=5 ttl=255 time=1 ms

  --- 192.168.30.1 ping statistics ---
    5 packet(s) transmitted
    5 packet(s) received
    0.00% packet loss
    round-trip min/avg/max = 1/1/1 ms
```

#提示：路由器RTC通过ping命令向IP为192.168.30.1的RTB发送5个ICMP包，收到了5个答复（Reply），每个包的字节数（bytes）都为56，序列号（Sequence）分别为1~5，生存时间（ttl）为255，耗费时间（time）为1ms，丢包率为0%，说明二者之间的网络已经互通，状态良好

```
[RTC]ping 192.168.40.1
  PING 192.168.40.1: 56  data bytes, press CTRL_C to break
    Reply from 192.168.40.1: bytes=56 Sequence=1 ttl=128 time=2 ms
    Reply from 192.168.40.1: bytes=56 Sequence=2 ttl=128 time=1 ms
    Reply from 192.168.40.1: bytes=56 Sequence=3 ttl=128 time=2 ms
    Reply from 192.168.40.1: bytes=56 Sequence=4 ttl=128 time=1 ms
    Reply from 192.168.40.1: bytes=56 Sequence=5 ttl=128 time=2 ms

  --- 192.168.40.1 ping statistics ---
    5 packet(s) transmitted
    5 packet(s) received
    0.00% packet loss
    round-trip min/avg/max = 1/1/2 ms
```

#提示：路由器RTC通过ping命令向IP为192.168.40.1的PC2发送5个ICMP包，收到了5个答复（Reply），每个包的字节数（bytes）都为56，序列号（Sequence）分别为1~5，生存时间（ttl）为128，耗费时间（time）为1ms或2ms，丢包率为0%，说明二者之间的网络已经互通，状态良好

```
[RTC]ping 192.168.20.1
  PING 192.168.20.1: 56  data bytes, press CTRL_C to break
    Request time out
    Request time out
    Request time out
```

```
        Request time out
        Request time out

    --- 192.168.20.1 ping statistics ---
      5 packet(s) transmitted
      0 packet(s) received
      100.00% packet loss
```
#提示：路由器RTC通过ping命令向IP地址为192.168.20.1的RTA发送5个ICMP包，数据包的字节数(bytes)为56字节，却未在规定时间内收到答复(Reply)，系统反馈5个请求包超时(Request time out)，丢包率为100%，说明二者之间的网络未连通。

```
    [RTC]ping 192.168.10.1
      PING 192.168.10.1: 56  data bytes, press CTRL_C to break
        Request time out
        Request time out
        Request time out
        Request time out
        Request time out

    --- 192.168.10.1 ping statistics ---
      5 packet(s) transmitted
      0 packet(s) received
      100.00% packet loss

    [RTC]
```
#提示：路由器RTC通过ping命令向IP地址为192.168.10.1的PC1发送5个ICMP包，数据包的字节数(bytes)为56，却未在规定时间内收到答复(Reply)，系统反馈5个请求包超时(Request time out)，丢包率为100%，说明二者的网络未连通。

9. 为了能够访问路由器 RTC 和 PC2，需要在路由器 RTA 上配置去往 192.168.30.0/24 和 192.168.40.0/24 的路由。

```
    [RTA]ip route-static 192.168.30.0 255.255.255.0 192.168.20.2   //配置去往192.168.30.0/24
的路由，其中192.168.20.2为数据包的下一跳地址
    [RTA]ip route-static 192.168.40.0 255.255.255.0 192.168.20.2   //配置去往192.168.40.0/24
的路由，其中192.168.20.2为数据包的下一跳地址
    [RTA]quit
    <RTA>save
```

小贴士

在路由器上配置静态路由，既可以为目标网络指定下一跳地址，也可以为目标网络指定输出接口。例如，在 RTA 上配置去往 192.168.30.0/24 的路由也可以写为：

```
    Ip route-static 192.168.30.0 255.255.255.0 g0/0/2    //指定去往192.168.30.0网络的数据包由
本地接口GE0/0/2转发
```

10. 查看路由器 RTA 的路由表状态时，能够看到系统增加了两条协议为 Static 的路由，Flags 为 RD，表示配置时使用的命令为下一跳地址指向的配置方式。

```
    [RTA]display ip routing-table
      Route Flags: R - relay, D - download to fib
    ------------------------------------------------------------
      Routing Tables: Public
```

```
        Destinations : 12       Routes : 12

Destination/Mask    Proto  Pre  Cost  Flags  NextHop          Interface
...
192.168.10.0/24     Direct 0    0     D      192.168.10.254   GE0/0/1
192.168.10.254/32   Direct 0    0     D      127.0.0.1        GE0/0/1
192.168.10.255/32   Direct 0    0     D      127.0.0.1        GE0/0/1
192.168.20.0/24     Direct 0    0     D      192.168.20.1     GE0/0/2
192.168.20.1/32     Direct 0    0     D      127.0.0.1        GE0/0/2
192.168.20.255/32   Direct 0    0     D      127.0.0.1        GE0/0/2
192.168.30.0/24     Static 60   0     RD     192.168.20.2     GE0/0/2
192.168.40.0/24     Static 60   0     RD     192.168.20.2     GE0/0/2
255.255.255.255/32  Direct 0    0     D      127.0.0.1        InLoopback0
```

小贴士

路由表中的每一行均表示一个目的网络的路由，也叫一个路由条目（Entries），路由器 RTA 的路由表中显示去往网络 192.168.30.0 的下一跳（FlagsNextHop）的地址为 192.168.20.2，输出接口（Interface）为 GE0/0/2。

11. 在 RTB 上配置去往 192.168.10.0/24、192.168.40.0/24 的静态路由，以实现路由器 RTB 与 PC1 与 PC2 的互访。

```
<RTB>system-view
Enter system view, return user view with Ctrl+Z.
[RTB]ip route-static 192.168.10.0 24 192.168.20.1    //配置去往192.168.10.0/24的路由
[RTB]ip route-static 192.168.40.0 24 192.168.30.2    //配置去往192.168.40.0/24的路由
[RTB] quit        //退出系统视图
<RTB>save         //保存配置
```

12. 查看路由器 RTB 的路由表状态，可以看到两条协议为 Static 的路由，分别指向 192.168.10.0/24 和 192.168.40.0/24 这两个目标网络，下一跳地址分别为 192.168.20.1 和 192.168.30.2。

```
[RTB]display ip routing-table
Route Flags: R - relay, D - download to fib
------------------------------------------------------------------------
Routing Tables: Public
        Destinations : 12       Routes : 12

Destination/Mask    Proto  Pre  Cost  Flags  NextHop          Interface
...                 //省略本地环回接口及路由信息
192.168.10.0/24     Static 60   0     RD     192.168.20.1     GE0/0/1
192.168.20.0/24     Direct 0    0     D      192.168.20.2     GE0/0/1
192.168.20.2/32     Direct 0    0     D      127.0.0.1        GE0/0/1
192.168.20.255/32   Direct 0    0     D      127.0.0.1        GE0/0/1
192.168.30.0/24     Direct 0    0     D      192.168.30.1     GE0/0/2
192.168.30.1/32     Direct 0    0     D      127.0.0.1        GE0/0/2
192.168.30.255/32   Direct 0    0     D      127.0.0.1        GE0/0/2
192.168.40.0/24     Static 60   0     RD     192.168.30.2     GE0/0/2
255.255.255.255/32  Direct 0    0     D      127.0.0.1        InLoopback0
```
#提示：RTB路由表中当前显示出两条协议（Proto）为静态（Static）的路由，分别去往192.168.10.0和

192.168.40.0，下一跳（FlagsNextHop）地址分别为192.168.20.1和192.168.30.2，输出接口为GE0/0/1和GE0/0/2

    [RTB]

13. 在路由器 RTC 上配置去往 192.168.10.0/24、192.168.20.0/24 的静态路由，以实现路由器 RTC 与 PC1 及路由器 RTA 的网络连通。

```
<RTC>system-view
Enter system view, return user view with Ctrl+Z.
[RTC]ip route-static 192.168.10.0  24 192.168.30.1    //配置去往192.168.10.0/24的路由
[RTC]ip route-static 192.168.20.0 24 192.168.30.1     //配置去往192.168.20.0/24的路由
[RTC]quit
<RTC>save
```

14. 查看路由器 RTC 的接口状态和路由表状态。

```
[RTC]display ip routing-table
Route Flags: R - relay, D - download to fib
------------------------------------------------------------
Routing Tables: Public
         Destinations : 12      Routes : 12

Destination/Mask    Proto   Pre  Cost   Flags NextHop         Interface
...
192.168.10.0/24     Static  60   0      RD    192.168.30.1    GE0/0/2
192.168.20.0/24     Static  60   0      RD    192.168.30.1    GE0/0/2
192.168.30.0/24     Direct  0    0      D     192.168.30.2    GE0/0/2
192.168.30.2/32     Direct  0    0      D     127.0.0.1       GE0/0/2
192.168.30.255/32   Direct  0    0      D     127.0.0.1       GE0/0/2
192.168.40.0/24     Direct  0    0      D     192.168.40.254  GE0/0/1
192.168.40.254/32   Direct  0    0      D     127.0.0.1       GE0/0/1
192.168.40.255/32   Direct  0    0      D     127.0.0.1       GE0/0/1
255.255.255.255/32  Direct  0    0      D     127.0.0.1       InLoopback0
#提示：RTC中已出现去往192.168.10.0和192.168.20.0的路由
[RTC]
```

15. 测试连通性，PC1 ping PC2，收到回复包，说明网络已连通，如图3-6-4所示。

图3-6-4　连通性测试

任务总结与思考

本任务重点讲述了使用路由器实现本地网络与远程网络互访的配置方法，同时介绍了静态路由的特点。

思考以下两个问题。

1. 当本地路由器上配置了去往目标路由器的路由时，如果没有在远程路由器上配置返回本地网络的路由，则会出现什么情况？网络能够连通吗？

2. 配置静态路由时使用下一跳地址或使用输出接口均可实现通信，两种配置方法在路由器上的执行效率是否相同？为什么？

知识补给

每个运行中的路由器都有一个路由表，路由器转发数据包的关键是路由表。在如图 3-6-5 所示的路由表中，每条路由表项都指明了数据包要到达某网络或某主机时应通过路由器的哪个物理接口发送，以及可到达该路径的哪个下一跳路由器，或者不再经过其他的路由器而直接可以到达的目的地。

```
[Huawei]display ip routing-table
Route Flags: R - relay, D - download to fib
------------------------------------------------------------
Routing Tables: Public   Destinations : 2      Routes : 2
Destination/Mask   Proto   Pre  Cost  Flags  NextHop    Interface

0.0.0.0/0          Static  60   0      D     120.0.0.2  Serial1/0/0
8.0.0.0/8          RIP     100  3      D     120.0.0.2  Serial1/0/0
9.0.0.0/8          OSPF    10   50     D     20.0.0.2   Ethernet2/0/0
9.1.0.0/16         RIP     100  4      D     120.0.0.2  Serial1/0/0
11.0.0.0/8         Static  60   0      D     120.0.0.2  Serial2/0/0
20.0.0.0/8         Direct  0    0      D     20.0.0.1   Ethernet2/0/0
20.0.0.1/32        Direct  0    0      D     127.0.0.1  LoopBack0
```

图 3-6-5　路由表

路由表中包含了下列关键项。

目的地址（Destination）：用来标识 IP 数据包的目的地址或目的网络。

网络掩码（Mask）：在 IP 编址课程中已经介绍了网络掩码的结构和作用。同样，在路由表中，网络掩码也具有重要的意义。IP 地址与网络掩码进行"逻辑与"便可得到相应的网段信息。例如，本例中的目的地址为 8.0.0.0、掩码为 255.0.0.0，"逻辑与"后便可得到一个 A 类网段信息（8.0.0.0/8）。网络掩码的作用还表现在，当路由表中有多条目的地址相同的路由信息时，路由器将选择其掩码最长的一项作为匹配项。

协议（Protocol）：路由条目的来源途径，其中，Direct 是指路由器直连接口形成的路由，Static 是指管理员手工配置的静态路由，OSPF、RIP 代表通过某种动态路由协议获取的路由。

优先级（Preference）：表明该路由的优先级，优先级的指定往往与其来源直接相关。

下一跳 IP 地址（NextHop）：指明 IP 数据包所经由的下一跳路由器的接口地址。

输出接口（Interface）：指明 IP 数据包将从该路由器的哪个接口转发出去。

任务拓展

在本任务的情景与拓扑结构都不变化的情况下，请同学们尝试用指定输出接口的方法为路由器配置静态路由，实现本地网络与外部网络的互访，比较本任务中的路由器路由表，查看有什么不同。

小技巧

在图 3-6-4 中，在 PC 上使用 tracert 命令，可以查看从当前设备到达目标设备所经过的所有设备的 IP 地址。从显示结果中可以看到，PC1 数据包到达 PC2 所经历的设备分别为 192.168.10.254（本地网关）、192.168.20.2（RTB 的 GE0/0/1 接口）、192.168.30.2（RTC 的 GE0/0/2 接口）。这条命令也可以运用在路由器上，例如：

```
 tracert  192.168.40.1
 tracert  192.168.40.1(192.168.40.1), max hops: 30 ,packet length: 40,
 press CTRL_C to break
1 192.168.10.254      1 ms     1 ms     1 ms
2 192.168.20.254      1 ms     1 ms     1 ms
3 192.168.30.2        2 ms     1 ms     1 ms
4 192.168.40.1        2 ms     1 ms     1 ms
```

命令的回显信息可以用来验证数据包的走向。

思考与实训 3

一、填空题

1. HUAWEIS5720-36PC-EI 对应 OSI 模型中的_____层设备。
2. 交换机之间互连的接口一般应配置为_____模式。
3. 若要在交换机上批量创建 VLAN11～VLAN15、VLAN20、VLAN30，则应使用命令_____。
4. 路由器属于 OSI 模型中_____层对应的设备。

5．可以给路由器供电的电源有_____、_____两种类型。

6．HUAWEIAR2220E 路由器背板上预留了_____个 SIC 槽位和_____个 WSIC 槽位。

7．路由器属于 OSI 模型中的_____层设备，通常包含_____、_____类型的接口。

8．静态路由通常是指由_____产生的路由。

9．静态路由的配置命令中必须包含_____、_____两项内容，以共同确定准确的目标网络。

10．静态路由配置方法中若使用 nexthop 或 interface，则执行效率更高的是_____。

二、判断题

1．每台交换机出厂时都已经创建了一个默认 VLAN，此 VLAN 也称管理 VLAN。（   ）

2．华为交换机默认已经开启了 VLANIF 接口，并在此基础上实现 VLAN 的三层互访。（   ）

3．路由器的性能优劣是由其硬件版本决定的。（   ）

4．路由器进行系统升级后，原有配置及密码都会丢失。（   ）

5．路由器启动后，首先进行硬件自检，然后才开始加载操作系统。（   ）

6．和 TFTP 相比，FTP 的灵活性更好，但可靠性稍差。（   ）

7．路由器的带内管理是指通过网络进行远程管理，所以带就是指"网络"。（   ）

8．HUAWEIAR2200E 路由器出厂时设置了一个默认 IP，不能修改。（   ）

9．HUAWEIAR220E 路由器默认提供 FTP 服务，连接 PC 后可以下载文件到路由器。（   ）

三、实训题

1．按照任务 2 中路由器的安装步骤，动手安装一台 HUAWEIAR2220E 路由器，并将一块 WSIC 单板接口安装到对应的槽位中。

2．自己动手，连接路由器和 PC，连接成功后，修改路由器主机名。

3．按照任务 3 的拓扑图及情景要求，利用单臂路由实现两个以上 VLAN 的互通与互访。

项目 4

动态管理路由

☆ 项目背景

宇信公司网络规划与建设项目进展顺利,已经完成了市场部、财务部、行政部等部门的业务网络部署和物理互连,各部门之间需要实现有序互访。由于网段比较多,因此高工程师建议配置动态路由协议,以实现全网动态感知路由,并采用协议技术对网络通信进行优化。

4.1 认识动态路由协议 RIP

认识动态路由协议 RIP

➢ **任务情景**

静态路由虽然可以实现互访，但是由于部门较多，网段较多，人工添加、管理、更新路由的工作量非常大，并且容易出错。为了能够让整个网络自动实现路由学习、更新，高工程师建议在整个网络上部署动态路由协议。小李选择部署简单、运行高效的 RIP 来实现。

➢ **任务分析**

- ➢ 了解 RIP 的基本功能；
- ➢ 学会在路由器上配置 RIP；
- ➢ 能看懂路由表中的 RIP。

➢ **实施准备**

1. HUAWEIAR2220E 路由器 3 台；
2. 双绞线 2 根；
3. Console 通信线缆 1 根；
4. 调测用 PC 1 台（预装 CRT 软件）。

➢ **实施步骤**

1. 按照图 4-1-1 所示 RIP 配置拓扑图连接网络设备。

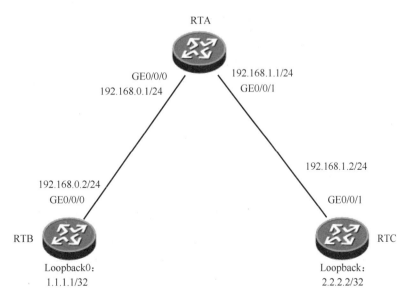

图 4-1-1　RIP 配置拓扑图

2. 配置路由器的接口、IP 地址和子网掩码（见表 4-1-1）。

表 4-1-1 设备地址分配表

设备名称	接口	IP 地址	子网掩码(Mask)
RTA	GE0/0/0	192.168.0.1	255.255.255.0
	GE0/0/1	192.168.1.1	255.255.255.0
RTB	GE0/0/0	192.168.0.2	255.255.255.0
	Loopback0	1.1.1.1	255.255.255.255
RTC	GE0/0/1	192.168.1.2	255.255.255.0
	Loopback0	2.2.2.2	255.255.255.255

3．在 3 台路由器上分别配置主机名称 RTA、RTB、RTC。

第一台路由器：

```
<Huawei>system-view        //进入系统视图
Enter system view, return user view with Ctrl+Z.
[Huawei]sysname RTA        //设置第一台路由器设备名称为RTA
[RTA]
```

第二台路由器：

```
<Huawei>system-view        //进入配置模式
Enter system view, return user view with Ctrl+Z.
[Huawei]sysname RTB        //设置第二台路由器设备名称为RTB
[RTB]
```

第三台路由器：

```
<Huawei>system-view        //进入配置模式
Enter system view, return user view with Ctrl+Z.
[Huawei]sysname RTC        //设置第三台路由器设备名称为RTC
[RTC]
```

4．在 RTA、RTB 和 RTC 上添加接口规划的地址。

```
[RTA]interface GigabitEthernet 0/0/0  //进入RTA设备的GE0/0/0接口
[RTA-GigabitEthernet0/0/0]ip address 192.168.0.1 255.255.255.0
//设置接口地址和子网掩码
Dec 24 2018 10:21:07-08:00 RTA %%01IFNET/4/LINK_STATE(l)[2]:The line protocol IP
 on the interface GigabitEthernet0/0/0 has entered the UP state.
[RTA-GigabitEthernet0/0/0]quit        //返回系统视图
[RTA]interface GigabitEthernet 0/0/1  //进入RTA设备的GE0/0/1接口
[RTA-GigabitEthernet0/0/1]ip address 192.168.1.1 255.255.255.0
//设置接口地址和子网掩码
Dec 24 2018 10:21:27-08:00 RTA %%01IFNET/4/LINK_STATE(l)[3]:The line protocol IP
 on the interface GigabitEthernet0/0/1 has entered the UP state.
[RTA-GigabitEthernet0/0/1]quit

[RTB]interface GigabitEthernet 0/0/0  //进入RTB设备的GE0/0/0接口
[RTB-GigabitEthernet0/0/0]ip address 192.168.0.2 255.255.255.0
//设置接口地址和子网掩码
Dec 24 2018 10:31:49-08:00 RTB %%01IFNET/4/LINK_STATE(l)[0]:The line protocol IP
 on the interface GigabitEthernet0/0/0 has entered the UP state.
```

```
[RTB-GigabitEthernet0/0/0]
[RTB-GigabitEthernet0/0/0]quit     //返回系统视图
[RTB]interface LoopBack 0  //进入RTB的Loopback0接口
[RTB-LoopBack0]ip address 1.1.1.1 255.255.255.255
//设置接口IP地址和子网掩码
[RTB-LoopBack0]quit     //返回系统视图

[RTC]interface GigabitEthernet 0/0/1       //进入RTC设备的GE0/0/0接口
[RTC-GigabitEthernet0/0/1]ip address 192.168.1.2 255.255.255.0
//设置接口地址和子网掩码
 Dec 24 2018 10:31:49-08:00 RTB %%01IFNET/4/LINK_STATE(l)[0]:The line protocol IP
 on the interface GigabitEthernet0/0/0 has entered the UP state.
[RTC-GigabitEthernet0/0/1]
[RTC-GigabitEthernet0/0/1]quit      //返回系统视图
[RTC]interface LoopBack 0  //进入RTC的Loopback0接口
[RTC-LoopBack0]ip address 2.2.2.2 255.255.255.255
//设置接口地址和子网掩码
[RTC-LoopBack0]quit
```

5. 在 RTA、RTB 和 RTC 设备上启动 RIPv2，并将接口地址网段发布到 RIPv2 中。

```
[RTA]rip                            //设备上启动RIP
[RTA-rip-1]version 2                //设置RIP的版本为V2版本
[RTA-rip-1]network 192.168.0.0      //RIP中发布GE0/0/0接口网段
[RTA-rip-1]network 192.168.1.0      //RIP中发布GE0/0/1接口网段
[RTA-rip-1]quit                     //返回系统视图
[RTA]

[RTB]rip                            //在设备上启动RIP
[RTB-rip-1]version 2                //设置RIP的版本为V2版本
[RTB-rip-1]network 192.168.0.0      //RIP中发布GE0/0/0接口网段
[RTB-rip-1]network 1.0.0.0          //RIP中发布Loopback0接口网段
[RTB-rip-1]quit                     //返回系统视图
[RTB]

[RTC]rip                            //在设备上启动RIP
[RTC-rip-1]version 2                //设置RIP的版本为V2版本
[RTC-rip-1]network 192.168.1.0      //RIP中发布GE0/0/0接口网段
[RTC-rip-1]network 2.0.0.0          //RIP中发布Loopback0接口网段
[RTC-rip-1]quit                     //退出RIP视图，返回系统视图
[RTC]
```

6. 在 RTA、RTB 和 RTC 的互连接口上启用 RIPv2 接口报文认证，并保存配置。

```
[RTA]interface GigabitEthernet 0/0/0
[RTA-GigabitEthernet0/0/0]rip authentication-mode ?   //根据系统提示设置认证模式
  hmac-sha256
  md5         MD5 authentication//MD5认证
  simple      Simple authentication  //密码认证
[RTA-GigabitEthernet0/0/0]rip authentication-mode simple ?//根据提示设置密码显示方式
  STRING<1-16>/<24,32>  Plain text/Encrypted text
  cipher              Encryption type (Cryptogram)     //密文显示
  plain               Encryption type (Plain text)     //明文显示
[RTA-GigabitEthernet0/0/0]rip authentication-mode simple plain huawei
//设置密码为huawei
```

```
            [RTA-GigabitEthernet0/0/0]quit
        [RTA]interface GigabitEthernet 0/0/1
        [RTA-GigabitEthernet0/0/1]rip authentication-mode md5 usual plain yuxin//G0/0/1
//设置认证方式为MD5加密认证，密钥为yuxin，usual代表MD5进行密文验证报文时，采用互联网通用的报文格
式（IETF标准）
        [RTA-GigabitEthernet0/0/1]
        [RTA-GigabitEthernet0/0/1]quit//返回系统视图
        [RTA]quit
        <RTA>save //保存配置
          The current configuration will be written to the device.
        Are you sure to continue? (y/n)[n]:y //输入yes，确认保存配置
        #提示：修改配置对于设备来说是一个危险操作，系统要求再次进行确认
          It will take several minutes to save configuration file, please wait........
          Configuration file had been saved successfully
          Note: The configuration file will take effect after being activated

        [RTB]interface GigabitEthernet 0/0/0
        [RTB-GigabitEthernet0/0/0]rip authentication-mode simple plain huawei
//设置接口RIP认证方式为密码认证，密码为明文显示huawei
        [RTB-GigabitEthernet0/0/0]quit //返回系统视图
        [RTB]quit
        <RTB>save//保存配置
          The current configuration will be written to the device.
        Are you sure to continue? (y/n)[n]:y//输入yes，确认保存配置
        #提示：修改配置对于设备来说是一个危险操作，系统要求再次进行确认
          It will take several minutes to save configuration file, please wait........
          Configuration file had been saved successfully
          Note: The configuration file will take effect after being activated
        <RTB>

        [RTC]interface GigabitEthernet 0/0/1
        [RTC-GigabitEthernet0/0/1]rip authentication-mode md5 usual plain yuxin
//设置GE0/0/1接口的RIP认证方式为MD5，密钥为yuxin，usual代表进行MD5密文验证报文时，采用互联网通
用报文格式（IETF标准）
        [RTC-GigabitEthernet0/0/1]quit    //返回系统视图
        [RTC]quit                         //返回用户视图
        <RTC>save                         //保存配置
          The current configuration will be written to the device.
        Are you sure to continue? (y/n)[n]:y//输入yes，确认保存配置
        #提示：修改配置对于设备来说是一个危险操作，系统要求再次进行确认
          It will take several minutes to save configuration file, please wait........
          Configuration file had been saved successfully
          Note: The configuration file will take effect after being activated
        <RTC>
```

7. 通过 display 命令查看 3 台路由器的路由表，以确认配置是否成功。

```
        [RTA]display ip routing-table protocol rip//查看通过RIP学习到的路由
        Route Flags: R - relay, D - download to fib
        ----------------------------------------------------------------
        Public routing table : RIP
                 Destinations : 2        Routes : 2

        RIP routing table status : <Active>
```

```
        Destinations : 2        Routes : 2

Destination/Mask    Proto   Pre   Cost   Flags   NextHop        Interface

1.1.1.1/32          RIP     100   1      D       192.168.0.2    GigabitEthernet0/0/0
2.2.2.2/32          RIP     100   1      D       192.168.1.2    GigabitEthernet0/0/1

RIP routing table status : <Inactive>
        Destinations : 0        Routes : 0

[RTA]display ip routing-table //查看全局路由表的路由信息
Route Flags: R - relay, D - download to fib
------------------------------------------------------------------
Routing Tables: Public
        Destinations : 12       Routes : 12

Destination/Mask     Proto    Pre   Cost   Flags   NextHop        Interface

1. 1.1.1/32          RIP      100   1      D       192.168.0.2    GigabitEthernet0/0/0
2. 2.2.2/32          RIP      100   1      D       192.168.1.2    GigabitEthernet0/0/1
127.0.0.0/8          Direct   0     0      D       127.0.0.1      InLoopBack0
127.0.0.1/32         Direct   0     0      D       127.0.0.1      InLoopBack0
127.255.255.255/32   Direct   0     0      D       127.0.0.1      InLoopBack0
192.168.0.0/24       Direct   0     0      D       192.168.0.1    GigabitEthernet0/0/0
192.168.0.1/32       Direct   0     0      D       127.0.0.1      GigabitEthernet0/0/0
192.168.0.255/32     Direct   0     0      D       127.0.0.1      GigabitEthernet0/0/0
192.168.1.0/24       Direct   0     0      D       192.168.1.1    GigabitEthernet0/0/1
192.168.1.1/32       Direct   0     0      D       127.0.0.1      GigabitEthernet0/0/1
192.168.1.255/32     Direct   0     0      D       127.0.0.1      GigabitEthernet0/0/1
255.255.255.255/32   Direct   0     0      D       127.0.0.1      InLoopBack0
```

#备注：以上为路由表信息，RTA学习到两条RIP路由，以1.1.1.1/32这条路由为例，该路由的目的地址和掩码为1.1.1.1/32，路由来源是RIP，优先级为100，cost值为1，数据包的下一跳地址为192.168.0.2，送出接口为GE0/0/0

```
[RTB]display ip routing-table protocol rip //查看RIP学习到的路由
Route Flags: R - relay, D - download to fib
------------------------------------------------------------------
Public routing table : RIP
        Destinations : 2        Routes : 2

RIP routing table status : <Active>
        Destinations : 2        Routes : 2

Destination/Mask    Proto   Pre   Cost   Flags   NextHop        Interface

  2.2.2.2/32        RIP     100   2      D       192.168.0.1    GigabitEthernet0/0/0
  192.168.1.0/24    RIP     100   1      D       192.168.0.1    GigabitEthernet0/0/0

RIP routing table status : <Inactive>
        Destinations : 0        Routes : 0

[RTB]display ip routing-table //查看全局路由表的路由信息
Route Flags: R - relay, D - download to fib
------------------------------------------------------------------
Routing Tables: Public
```

```
        Destinations : 10       Routes : 10

Destination/Mask    Proto   Pre  Cost  Flags  NextHop       Interface

1.1.1.1/32          Direct  0    0     D      127.0.0.1     Loopback0
2.2.2.2/32          RIP     100  2     D      192.168.0.1   GigabitEthernet0/0/0
127.0.0.0/8         Direct  0    0     D      127.0.0.1     InLoopback0
127.0.0.1/32        Direct  0    0     D      127.0.0.1     InLoopback0
127.255.255.255/32  Direct  0    0     D      127.0.0.1     InLoopback0
192.168.0.0/24      Direct  0    0     D      192.168.0.2   GigabitEthernet0/0/0
192.168.0.2/32      Direct  0    0     D      127.0.0.1     GigabitEthernet0/0/0
192.168.0.255/32    Direct  0    0     D      127.0.0.1     GigabitEthernet0/0/0
192.168.1.0/24      RIP     100  1     D      192.168.0.1   GigabitEthernet0/0/0
255.255.255.255/32  Direct  0    0     D      127.0.0.1     InLoopback0
```

#提示：以上为路由表信息，RTB学习到两条RIP路由，以2.2.2.2/32这条路由，该路由的目的地址和掩码为2.2.2.2/32，路由来源是RIP，优先级为100，cost为2，数据包的下一跳地址为192.168.0.1，路由器转发接口（送出接口）为GE0/0/0

```
[RTC]display ip routing-table protocol rip //查看RIP学习到的路由
Route Flags: R - relay, D - download to fib
------------------------------------------------------------------
Public routing table : RIP
        Destinations : 2        Routes : 2

RIP routing table status : <Active>
        Destinations : 2        Routes : 2

Destination/Mask  Proto  Pre  Cost  Flags  NextHop       Interface

1.1.1.1/32        RIP    100  2     D      192.168.1.1   GigabitEthernet0/0/1
192.168.0.0/24    RIP    100  1     D      192.168.1.1   GigabitEthernet0/0/1

[RTC]display ip routing-table //查看全局路由表的路由信息
Route Flags: R - relay, D - download to fib
------------------------------------------------------------------
Routing Tables: Public
        Destinations : 10       Routes : 10

Destination/Mask    Proto   Pre  Cost  Flags  NextHop       Interface

1.1.1.1/32          RIP     100  2     D      192.168.1.1   GigabitEthernet0/0/1
2.2.2.2/32          Direct  0    0     D      127.0.0.1     Loopback0
127.0.0.0/8         Direct  0    0     D      127.0.0.1     InLoopback0
127.0.0.1/32        Direct  0    0     D      127.0.0.1     InLoopback0
127.255.255.255/32  Direct  0    0     D      127.0.0.1     InLoopback0
192.168.0.0/24      RIP     100  1     D      192.168.1.1   GigabitEthernet0/0/1
192.168.1.0/24      Direct  0    0     D      192.168.1.2   GigabitEthernet0/0/1
192.168.1.2/32      Direct  0    0     D      127.0.0.1     GigabitEthernet0/0/1
192.168.1.255/32    Direct  0    0     D      127.0.0.1     GigabitEthernet0/0/1
255.255.255.255/32  Direct  0    0     D      127.0.0.1     InLoopback0
```

#提示：以上为路由表信息，RTC学习到两条路由，以1.1.1.1/32这条路由为例，该路由的目的地址和掩码为1.1.1.1/32，路由来源是RIP，优先级为100，cost为2，数据包的下一跳地址为192.168.1.1，送出接口为GE0/0/1

8. 用ping命令测试RTB和RTC之间的连通性。

```
[RTB]ping -a 1.1.1.1 2.2.2.2    //以本设备的Loopback接口地址作为源地址ping RTC的2.2.2.2地址
  PING 2.2.2.2: 56  data bytes, press CTRL_C to break
    Reply from 2.2.2.2: bytes=56 Sequence=1 ttl=254 time=20 ms
    Reply from 2.2.2.2: bytes=56 Sequence=2 ttl=254 time=20 ms
    Reply from 2.2.2.2: bytes=56 Sequence=3 ttl=254 time=20 ms
    Reply from 2.2.2.2: bytes=56 Sequence=4 ttl=254 time=30 ms
    Reply from 2.2.2.2: bytes=56 Sequence=5 ttl=254 time=20 ms

  --- 2.2.2.2 ping statistics ---
    5 packet(s) transmitted
    5 packet(s) received
    0.00% packet loss
    round-trip min/avg/max = 20/22/30 ms

[RTC]ping -a 2.2.2.2 1.1.1.1    //以本设备的Loopback接口地址作为源地址ping RTB
  PING 1.1.1.1: 56  data bytes, press CTRL_C to break
    Reply from 1.1.1.1: bytes=56 Sequence=1 ttl=254 time=20 ms
    Reply from 1.1.1.1: bytes=56 Sequence=2 ttl=254 time=30 ms
    Reply from 1.1.1.1: bytes=56 Sequence=3 ttl=254 time=20 ms
    Reply from 1.1.1.1: bytes=56 Sequence=4 ttl=254 time=20 ms
    Reply from 1.1.1.1: bytes=56 Sequence=5 ttl=254 time=20 ms

  --- 1.1.1.1 ping statistics ---
    5 packet(s) transmitted
    5 packet(s) received
    0.00% packet loss
    round-trip min/avg/max = 20/22/30 ms
```

#提示：ping -a参数可以携带指定源地址ping目的地址，默认情况设备的ping报文是以送出接口地址为源地址的，但是如果想手动设置源地址则可以通过ping -a 源ip 目的ip来实现，如ping -a 2.2.2.2 1.1.1.1，该命令的作用是在设备上以2.2.2.2作为源地址ping目的地址1.1.1.1，通过该命令可以看出，RTC的2.2.2.2和RTB的1.1.1.1互通，证明RIP路由学习正常。

 小贴士

（1）HUAWEIAR2220E 路由器上支持 RIPv1 和 RIPv2 两个版本，目前主流使用 RIPv2 版本，RIPv1 版本在实际组网环境中不建议使用，HUAWEIAR2220 系列路由器上默认的 RIP 版本是 RIPv1，所以在配置过程中要注意手动更改 RIP 版本。

（2）环回接口（Loopback）是设备的虚拟接口，相比于物理接口，它更稳定，只要路由器运行正常，环回接口状态就是 UP 状态。本任务中，我们通过 RTB 和 RTC 设备上的环回接口 Loopback0 来检测 RIPv2 在路由器间的路由学习情况。

任务总结与思考

本任务重点讲述了路由器上如何通过动态路由协议 RIP 实现路由信息的互相学习和传递，并且介绍了 RIPv2 接口报文认证的验证配置方法，如果遇到了简单、小型的网络拓扑，则可以考虑通过 RIPv2 来实现。

思考以下两个问题。

1. 如果 RTB 或 RTC 路由器上的网段比较多，则可以假设有很多 Loopback 接口，如果逐一通过 network 方式发布，则配置复杂，并且工作量大，那么有没有更好的发布方式？

2. 如果 RTC 的 Loopback0 接口被误删，那么网络中的 RIP 路由会发生什么变化？

 知识补给

RIP 作为一种比较"古老"的路由协议，在其发展过程中出现了两个版本，分别是早期实现的 RIPv1 和目前用到的 RIPv2。RIPv1 本身存在固有的缺陷和问题。

（1）RIPv1 以广播方式发送报文，目的 IP 地址为 255.255.255.255，这种报文的发送方式给网络造成了严重的带宽浪费。

（2）RIPv1 本身不支持认证，所以 RIPv1 网络中的安全威胁无警惕性和处理机制。

（3）RIPv1 是有类路由协议，发送路由更新信息时不携带任何子网掩码，可能会造成有些设备无法"路由"的情况发生，同时 RIPv1 不支持变长子网掩码 VLSM。

RIPv2 针对 RIPv1 的一些问题进行了改进。

（1）RIPv2 的报文发送方式采用组播方式，目的地址为固定组播地址 224.0.0.9。组播方式发送报文的好处是，可避免同一网络内因没有运行 RIP 协议的接口接收该报文，而无端地消耗资源及带宽；相比 RIPv1 的广播方式，RIPv2 的组播发送方式大大提高了网络的转发效率和带宽利用率。

（2）RIPv2 支持认证，可以在 RIPv2 网络中的路由器互连接口上启用 RIP 报文认证，早期的 RIPv2 只支持简单明文认证，安全性低，因为明文认证密码串可以很轻易地被截获。随着对 RIP 安全性的需求越来越高，RIPv2 引入了加密认证功能，开始时通过支持 MD5 认证（RFC 2082）来实现，后来通过支持 HMAC-SHA-1 认证（RFC 2082）进一步增强了安全性。HUAWEIAR2220 系列路由器能够支持以上提到的所有认证方式，通过 `rip authentication-mode` 命令来配置接口的认证方式和密码信息，通过 `undo rip authentication-mode` 命令来取消接口的所有验证。通过认证，路由器可以识别对方邻居路由器是否是一个正确合法的邻居，如果认证不通过，则路由器会丢弃从邻居发送过来的所有路由更新信息。RIPv2 的认证效果如图 4-1-2 所示。

（3）RIPv2 支持 VLSM 和 CIDR，在常用的动态路由协议中，只有 RIPv1 支持有类的路由协议，其他协议（OSPF、ISIS、BGP）都支持无类的动态路由协议。

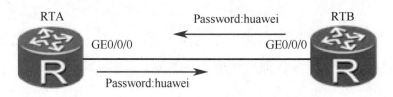

图 4-1-2　RIPv2 的认证效果

任务拓展

如图 4-1-3 所示，若在 RTA、RTB 和 RTC 路由器上运行 RIPv2，则 RTA 路由器上的路由网段比较多，如果逐个网段发布（通过 network 实现），则配置量大且效率低，有没有其他方式可以发布 RIPv2？路由引入是一个不错的选择，当在 RTA 上的 RIPv2 路由的网段比较多时，可以在 RIP 进程下执行 `import-route static/direct/`其他路由协议，该命令可以将本路由器上路由表中的静态、直连路由或其他路由协议的路由引入 RIPv2 中并发布出去。

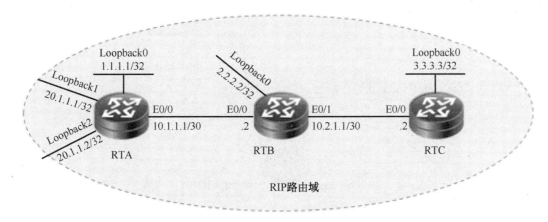

图 4-1-3　RIPv2 路由引入

小贴士

如果我们在 RIP 路由器上选择通过引入方式发布直连路由，则路由器间互连接口的地址一定要通过 network 发布。

小技巧

如图 4-1-3 所示，RTA 上的 Loopback1 和 Loopback2 都属于 20.1.1.0/30 网段，那么我们在用 network 发布该网段时只需要 network 20.1.1.0 即可，无须重复发布。

4.2　认识动态路由协议 OSPF（单区域）

> **任务情景**

宇信公司的各部门需要实现互访，虽然可以通过静态路由实现，但是由于部门比较多、网段比较多，静态路由在多网络互通场景中需要配置的路由条目很多，并且需要静态添加、删除和更新。为了实现动态的路由学习，小李决定选择动态路由协议 OSPF（单区域）来实现。

> **任务分析**

> 了解 OSPF 的基本功能；

- 学会在路由器上配置 OSPF；
- 能看懂路由表中的 OSPF 路由。

实施准备

1. HUAWEI AR2220E 路由器 3 台；
2. 网线 2 根；
3. Console 通信线缆 1 根；
4. 调测用 PC 1 台（预装 CRT 软件）。

实施步骤

1. 按照如图 4-2-1 所示连接网络设备。

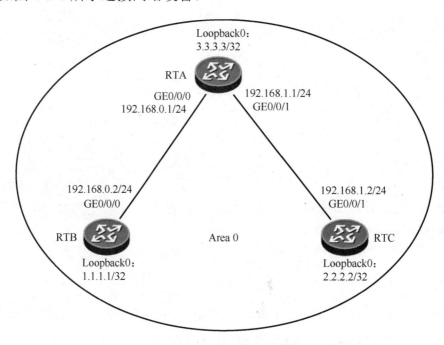

图 4-2-1 OSPF（单区域）配置拓扑图

2. 配置路由器的互连接口地址和子网掩码，设备地址分配如表 4-2-1 所示。

表 4-2-1 设备地址分配表

设备名称	接口	IP 地址	掩码(Mask)
RTA	GE0/0/0	192.168.0.1	255.255.255.0
	GE0/0/1	192.168.1.1	255.255.255.0
	Loopback0	3.3.3.3	255.255.255.255
RTB	GE0/0/0	192.168.0.2	255.255.255.0
	Loopback0	1.1.1.1	255.255.255.255
RTC	GE0/0/1	192.168.1.2	255.255.255.0
	Loopback0	2.2.2.2	255.255.255.255

3. 3台路由器上分别配置主机名称为RTA、RTB、RTC。

第一台路由器：

```
<Huawei>system-view          //进入配置模式
Enter system view, return user view with Ctrl+Z.
[Huawei]sysname RTA //设置第一台路由器设备名称为RTA
[RTA]
```

第二台路由器：

```
<Huawei>system-view          //进入配置模式
Enter system view, return user view with Ctrl+Z.
[Huawei]sysname RTB //设置第二台路由器设备名称为RTB
[RTB]
```

第三台路由器：

```
<Huawei>system-view          //进入配置模式
Enter system view, return user view with Ctrl+Z.
[Huawei]sysname RTC //设置第三台路由器设备名称为RTC
[RTC]
```

4. 在RTA、RTB和RTC上添加接口地址。

```
[RTA]interface GigabitEthernet 0/0/0    //进入RTA设备GE0/0/0接口
[RTA-GigabitEthernet0/0/0]ip address 192.168.0.1 255.255.255.0
//设置接口地址和子网掩码
Dec 24 2018 10:21:07-08:00 RTA %%01IFNET/4/LINK_STATE(l)[2]:The line protocol IP
 on the interface GigabitEthernet0/0/0 has entered the UP state.
[RTA-GigabitEthernet0/0/0]quit     //返回系统视图
[RTA]interface GigabitEthernet 0/0/1    //进入RTA设备GE0/0/1接口
[RTA-GigabitEthernet0/0/1]ip address 192.168.1.1 255.255.255.0
//设置接口IP地址和子网掩码
Dec 24 2018 10:21:27-08:00 RTA %%01IFNET/4/LINK_STATE(l)[3]:The line protocol IP
 on the interface GigabitEthernet0/0/1 has entered the UP state.
[RTA-GigabitEthernet0/0/1]quit
[RTA]interface LoopBack 0//进入RTA Loopback0接口
[RTA-LoopBack0]ip address 3.3.3.3 255.255.255.255//设置接口地址和子网掩码
[RTA-LoopBack0]quit    //返回系统视图
[RTA]
[RTB]interface GigabitEthernet 0/0/0      //进入RTB设备GE0/0/0接口
[RTB-GigabitEthernet0/0/0]ip address 192.168.0.2 255.255.255.0
//设置接口地址和子网掩码
Dec 24 2018 10:31:49-08:00 RTB %%01IFNET/4/LINK_STATE(l)[0]:The line protocol IP
 on the interface GigabitEthernet0/0/0 has entered the UP state.
[RTB-GigabitEthernet0/0/0]quit     //返回系统视图
[RTB]interface LoopBack 0     //进入RTB设备Loopback0接口
[RTB-LoopBack0]ip address 1.1.1.1 255.255.255.255//设置接口地址和子网掩码
[RTB-LoopBack0]quit    //返回系统视图

[RTC]interface GigabitEthernet 0/0/1       //进入RTC设备GE0/0/0接口
[RTC-GigabitEthernet0/0/1]ip address 192.168.1.2 255.255.255.0
//设置接口地址和子网掩码
Dec 24 2018 10:31:49-08:00 RTB %%01IFNET/4/LINK_STATE(l)[0]:The line protocol IP
 on the interface GigabitEthernet0/0/0 has entered the UP state.
[RTC-GigabitEthernet0/0/1]quit     //返回系统视图
```

```
[RTC]interface LoopBack 0        //进入RTB设备Loopback0接口
[RTC-LoopBack0]ip address 2.2.2.2 255.255.255.255//设置接口地址和子网掩码
[RTC-LoopBack0]quit
```

5. 启动 OSPF，将接口地址网段发布到 OSPF 的 area 0 中。

```
[RTA]router id 3.3.3.3   //设置系统OSPF协议的router id
Info: Router ID has been modified, please reset the relative protocols manually to update the Router ID.
[RTA]ospf       //启动OSPF，并进入协议视图
[RTA-ospf-1]area 0    //进入OSPF的骨干区域
#提示：OSPF路由协议中默认区域0，称为骨干区域
[RTA-ospf-1-area-0.0.0.0]network 192.168.0.0 0.0.0.255
//发布GE0/0/0接口所在的网络地址
[RTA-ospf-1-area-0.0.0.0]network 192.168.1.0 0.0.0.255
//发布GE0/0/1接口所在的网络地址
[RTA-ospf-1-area-0.0.0.0]network 3.3.3.3 0.0.0.0
//发布Lookback0接口所在的网络地址
[RTA-ospf-1-area-0.0.0.0]quit    //返回OSPF视图
[RTA-ospf-1]quit         //返回系统视图
[RTA]

[RTB]router id 1.1.1.1   //设置系统OSPF的router ID
Info: Router ID has been modified, please reset the relative protocols manually to update the Router ID.
[RTB]ospf       //启动OSPF，并进入协议视图
[RTB-ospf-1]area 0     //进入OSPF的骨干区域
[RTB-ospf-1-area-0.0.0.0]network 192.168.0.0 0.0.0.255
//发布GE0/0/0接口所在的网络地址
[RTB-ospf-1-area-0.0.0.0]network 1.1.1.1 0.0.0.0//发布Lookback0接口所在的网络地址
[RTB-ospf-1-area-0.0.0.0]quit     //返回OSPF视图
[RTB-ospf-1]quit      //退出并返回系统视图
[RTB]

[RTC]router id 2.2.2.2   //设置系统的OSPF的router ID
Info: Router ID has been modified, please reset the relative protocols manually to update the Router ID.
[RTC]ospf     //启动OSPF，并进入协议视图
[RTC-ospf-1]area 0      //进入OSPF的骨干区域
[RTC-ospf-1-area-0.0.0.0]network 192.168.1.0 0.0.0.255
//发布GE0/0/1接口所在的网络地址
[RTC-ospf-1-area-0.0.0.0]network 2.2.2.2 0.0.0.0
//发布Lookback0接口所在的网络地址
[RTC-ospf-1-area-0.0.0.0]quit    //返回OSPF视图
[RTC-ospf-1]quit    //退出并返回系统视图
[RTC]
```

6. 在 RTA、RTB 和 RTC 的互连接口上启用 OSPF 接口认证，并保存配置。

```
[RTA]interface GigabitEthernet 0/0/0
[RTA-GigabitEthernet0/0/0]ospf authentication-mode ?
  hmac-md5  Use HMAC-MD5 algorithm
  keychain  Keychain authentication mode
  md5       Use MD5 algorithm
```

```
    null    Use null authentication
    simple  Simple authentication mode
[RTA-GigabitEthernet0/0/0]ospf authentication-mode simple cipher huawei
[RTA-GigabitEthernet0/0/0]quit
[RTA]interface GigabitEthernet 0/0/1
[RTA-GigabitEthernet0/0/1]ospf authentication-mode simple cipher yuxin
[RTA-GigabitEthernet0/0/1]quit
[RTA]quit
<RTA>save  //保存配置
  The current configuration will be written to the device.
Are you sure to continue? (y/n)[n]:y //输入yes，确认保存配置
#提示：修改配置对于设备来说是一个危险操作，系统要求再次进行确认
  It will take several minutes to save configuration file, please wait.......
  Configuration file had been saved successfully
  Note: The configuration file will take effect after being activated

[RTB]interface GigabitEthernet 0/0/0
[RTB-GigabitEthernet0/0/0]ospf authentication-mode simple cipher huawei
[RTB-GigabitEthernet0/0/0]quit       //退出并返回系统视图
[RTB]quit
<RTB>save//保存配置
  The current configuration will be written to the device.
Are you sure to continue? (y/n)[n]:y//输入yes，确认保存配置
#提示：修改配置对于设备来说是一个危险操作，系统要求再次进行确认
  It will take several minutes to save configuration file, please wait........
  Configuration file had been saved successfully
  Note: The configuration file will take effect after being activated
<RTB>


[RTC]interface GigabitEthernet 0/0/1
[RTC-GigabitEthernet0/0/1]ospf authentication-mode simple cipher yuxin
[RTC-GigabitEthernet0/0/1]quit        //退出并返回系统视图
[RTC] quit         //退出并返回用户视图
<RTC>save //保存配置
  The current configuration will be written to the device.
Are you sure to continue? (y/n)[n]:y//输入yes，确认保存配置
#提示：修改配置对于设备来说是一个危险操作，系统要求再次进行确认
  It will take several minutes to save configuration file, please wait........
  Configuration file had been saved successfully
  Note: The configuration file will take effect after being activated
<RTC>
```

7. 查看三台路由器的 OSPF 邻居关系是否正常。

```
[RTA]display ospf peer brief

  OSPF Process 1 with Router ID 3.3.3.3
        Peer Statistic Information
 ----------------------------------------------------------------
  Area Id           Interface                Neighbor id        State
  0.0.0.0           GigabitEthernet0/0/0     1.1.1.1            Full
  0.0.0.0           GigabitEthernet0/0/1     2.2.2.2            Full
 ----------------------------------------------------------------
#提示：通过以上输出信息可以看出，RTA有2个邻居，分别是1.1.1.1和2.2.2.2，Full状态代表邻居关系正常
```

```
[RTB]display ospf peer brief

 OSPF Process 1 with Router ID 1.1.1.1
       Peer Statistic Information
 ----------------------------------------------------------------
 Area Id           Interface                Neighbor id       State
 0.0.0.0           GigabitEthernet0/0/0     3.3.3.3           Full
 ----------------------------------------------------------------
```

#提示：通过以上输出信息可以看出，RTB 有 1 个邻居，为 3.3.3.3，Full 状态代表邻居关系正常

```
[RTC]display ospf peer brief

 OSPF Process 1 with Router ID 2.2.2.2
       Peer Statistic Information
 ----------------------------------------------------------------
 Area Id           Interface                Neighbor id       State
 0.0.0.0           GigabitEthernet0/0/1     3.3.3.3           Full
 ----------------------------------------------------------------
```

#提示：通过以上输出信息可以看出，RTC 有 1 个邻居，为 3.3.3.3，Full 状态代表邻居关系正常

8. 通过 display 命令查看 3 台路由器上的 OSPF 路由是否学习成功，路由表是否正常。

查看 RTA 上的路由信息：

```
[RTA]display ip routing-table protocol ospf
Route Flags: R - relay, D - download to fib
------------------------------------------------------------------
Public routing table : OSPF
        Destinations : 2        Routes : 2

OSPF routing table status : <Active>
        Destinations : 2        Routes : 2

Destination/Mask   Proto   Pre   Cost   Flags  NextHop         Interface

  1.1.1.1/32       OSPF    10    1      D      192.168.0.2     GigabitEthernet0/0/0
  2.2.2.2/32       OSPF    10    1      D      192.168.1.2     GigabitEthernet0/0/1

OSPF routing table status : <Inactive>
        Destinations : 0        Routes : 0

[RTA]display ip routing-table
Route Flags: R - relay, D - download to fib
------------------------------------------------------------------
Routing Tables: Public
        Destinations : 13       Routes : 13

Destination/Mask   Proto    Pre   Cost   Flags  NextHop         Interface

  1.1.1.1/32       OSPF     10    1      D      192.168.0.2     GigabitEthernet0/0/0
  2.2.2.2/32       OSPF     10    1      D      192.168.1.2     GigabitEthernet0/0/1
  3.3.3.3/32       Direct   0     0      D      127.0.0.1       Loopback0
  127.0.0.0/8      Direct   0     0      D      127.0.0.1       InLoopback0
  127.0.0.1/32     Direct   0     0      D      127.0.0.1       InLoopback0
```

```
127.255.255.255/32  Direct  0    0    D    127.0.0.1      InLoopback0
192.168.0.0/24      Direct  0    0    D    192.168.0.1    GigabitEthernet0/0/0
192.168.0.1/32      Direct  0    0    D    127.0.0.1      GigabitEthernet0/0/0
192.168.0.255/32    Direct  0    0    D    127.0.0.1      GigabitEthernet0/0/0
192.168.1.0/24      Direct  0    0    D    192.168.1.1    GigabitEthernet0/0/1
192.168.1.1/32      Direct  0    0    D    127.0.0.1      GigabitEthernet0/0/1
192.168.1.255/32    Direct  0    0    D    127.0.0.1      GigabitEthernet0/0/1
255.255.255.255/32  Direct  0    0    D    127.0.0.1      InLoopback0
```

#提示：通过以上输出可以看出，RTA学习到2条OSPF路由。以1.1.1.1/32路由条目为例，该路由的目的地址和掩码为1.1.1.1/32，路由来源是OSPF，优先级为10，cost为1，下一跳地址为192.168.0.2，输出接口为GE0/0/0

查看RTB设备上的路由信息：

```
[RTB]display ip routing-table protocol ospf
Route Flags: R - relay, D - download to fib
------------------------------------------------------------------
Public routing table : OSPF
       Destinations : 3        Routes : 3

OSPF routing table status : <Active>
       Destinations : 3        Routes : 3

Destination/Mask    Proto  Pre  Cost  Flags NextHop      Interface

2.2.2.2/32          OSPF   10   2     D     192.168.0.1  GigabitEthernet0/0/0
3.3.3.3/32          OSPF   10   1     D     192.168.0.1  GigabitEthernet0/0/0
192.168.1.0/24      OSPF   10   2     D     192.168.0.1  GigabitEthernet0/0/0

OSPF routing table status : <Inactive>
       Destinations : 0        Routes : 0


[RTB]display ip routing-table
Route Flags: R - relay, D - download to fib
------------------------------------------------------------------
Routing Tables: Public
       Destinations : 11       Routes : 11

Destination/Mask    Proto   Pre  Cost  Flags NextHop      Interface

1.1.1.1/32          Direct  0    0     D     127.0.0.1    Loopbback0
2.2.2.2/32          OSPF    10   2     D     192.168.0.1  GigabitEthernet0/0/0
3.3.3.3/32          OSPF    10   1     D     192.168.0.1  GigabitEthernet0/0/0
127.0.0.0/8         Direct  0    0     D     127.0.0.1    InLoopback0
127.0.0.1/32        Direct  0    0     D     127.0.0.1    InLoopback0
127.255.255.255/32  Direct  0    0     D     127.0.0.1    InLoopback0
192.168.0.0/24      Direct  0    0     D     192.168.0.2  GigabitEthernet0/0/0
192.168.0.2/32      Direct  0    0     D     127.0.0.1    GigabitEthernet0/0/0
192.168.0.255/32    Direct  0    0     D     127.0.0.1    GigabitEthernet0/0/0
192.168.1.0/24      OSPF    10   2     D     192.168.0.1  GigabitEthernet0/0/0
255.255.255.255/32  Direct  0    0     D     127.0.0.1    InLoopback0
```

#提示：通过以上输出可以看出，RTA学习到3条OSPF路由。以2.2.2.2/32路由条目为例，该路由的目的地址和掩码为2.2.2.2/32，路由来源是OSPF，优先级为10，cost为2，下一跳地址为192.168.0.1，输出接口为GE0/0/0

查看RTC设备上的路由信息：

```
[RTC]display ip routing-table protocol ospf
Route Flags: R - relay, D - download to fib
```

```
         ----------------------------------------------------------------
         Public routing table : OSPF
                 Destinations : 3        Routes : 3

         OSPF routing table status : <Active>
                 Destinations : 3        Routes : 3

         Destination/Mask   Proto  Pre  Cost  Flags  NextHop      Interface

         1.1.1.1/32         OSPF   10   2     D      192.168.1.1  GigabitEthernet0/0/1
         3.3.3.3/32         OSPF   10   1     D      192.168.1.1  GigabitEthernet0/0/1
         192.168.0.0/24     OSPF   10   2     D      192.168.1.1  GigabitEthernet0/0/1

         OSPF routing table status : <Inactive>
                 Destinations : 0        Routes : 0


         [RTC]disp ip routing-table
         Route Flags: R - relay, D - download to fib
         ----------------------------------------------------------------
         Routing Tables: Public
                 Destinations : 11       Routes : 11

         Destination/Mask   Proto   Pre  Cost  Flags  NextHop      Interface

         1.1.1.1/32         OSPF    10   2     D      192.168.1.1  GigabitEthernet0/0/1
         2.2.2.2/32         Direct  0    0     D      127.0.0.1    Loopback0
         3.3.3.3/32         OSPF    10   1     D      192.168.1.1  GigabitEthernet0/0/1
         127.0.0.0/8        Direct  0    0     D      127.0.0.1    InLoopback0
         127.0.0.1/32       Direct  0    0     D      127.0.0.1    InLoopback0
         127.255.255.255/32 Direct  0    0     D      127.0.0.1    InLoopback0
         192.168.0.0/24     OSPF    10   2     D      192.168.1.1  GigabitEthernet0/0/1
         192.168.1.0/24     Direct  0    0     D      192.168.1.2  GigabitEthernet0/0/1
         192.168.1.2/32     Direct  0    0     D      127.0.0.1    GigabitEthernet0/0/1
         192.168.1.255/32   Direct  0    0     D      127.0.0.1    GigabitEthernet0/0/1
         255.255.255.255/32 Direct  0    0     D      127.0.0.1    InLoopback0
```

#提示：通过以上输出可以看出，RTA学习到3条OSPF路由。以1.1.1.1/32路由条目为例，该路由的目的地址和掩码为1.1.1.1/32，路由来源是OSPF，优先级为10，cost为2，下一跳地址为192.168.1.1，输出接口为GE0/0/1。

9. 利用ping命令测试RTA、RTB和RTC之间的网络连通性。

```
[RTA]ping 1.1.1.1
  PING 1.1.1.1: 56  data bytes, press CTRL_C to break
    Reply from 1.1.1.1: bytes=56 Sequence=1 ttl=255 time=60 ms
    Reply from 1.1.1.1: bytes=56 Sequence=2 ttl=255 time=30 ms
    Reply from 1.1.1.1: bytes=56 Sequence=3 ttl=255 time=30 ms
    Reply from 1.1.1.1: bytes=56 Sequence=4 ttl=255 time=20 ms
    Reply from 1.1.1.1: bytes=56 Sequence=5 ttl=255 time=30 ms

  --- 1.1.1.1 ping statistics ---
    5 packet(s) transmitted
    5 packet(s) received
    0.00% packet loss
    round-trip min/avg/max = 20/34/60 ms
[RTA]ping 2.2.2.2
  PING 2.2.2.2: 56  data bytes, press CTRL_C to break
    Reply from 2.2.2.2: bytes=56 Sequence=1 ttl=255 time=20 ms
    Reply from 2.2.2.2: bytes=56 Sequence=2 ttl=255 time=30 ms
```

```
    Reply from 2.2.2.2: bytes=56 Sequence=3 ttl=255 time=30 ms
    Reply from 2.2.2.2: bytes=56 Sequence=4 ttl=255 time=20 ms
    Reply from 2.2.2.2: bytes=56 Sequence=5 ttl=255 time=20 ms

  --- 2.2.2.2 ping statistics ---
    5 packet(s) transmitted
    5 packet(s) received
    0.00% packet loss
    round-trip min/avg/max = 20/24/30 ms
[RTB]ping 2.2.2.2
PING 2.2.2.2: 56  data bytes, press CTRL_C to break
Reply from 2.2.2.2: bytes=56 Sequence=1 ttl=254 time=40 ms
Reply from 2.2.2.2: bytes=56 Sequence=2 ttl=254 time=30 ms
Reply from 2.2.2.2: bytes=56 Sequence=3 ttl=254 time=30 ms
Reply from 2.2.2.2: bytes=56 Sequence=4 ttl=254 time=50 ms
Reply from 2.2.2.2: bytes=56 Sequence=5 ttl=254 time=50 ms

--- 2.2.2.2 ping statistics ---
 5 packet(s) transmitted
 5 packet(s) received
 0.00% packet loss
 round-trip min/avg/max = 30/40/50 ms
[RTB]ping 3.3.3.3
PING 3.3.3.3: 56  data bytes, press CTRL_C to break
Reply from 3.3.3.3: bytes=56 Sequence=1 ttl=255 time=20 ms
Reply from 3.3.3.3: bytes=56 Sequence=2 ttl=255 time=20 ms
Reply from 3.3.3.3: bytes=56 Sequence=3 ttl=255 time=20 ms
Reply from 3.3.3.3: bytes=56 Sequence=4 ttl=255 time=20 ms
Reply from 3.3.3.3: bytes=56 Sequence=5 ttl=255 time=20 ms

--- 3.3.3.3 ping statistics ---
 5 packet(s) transmitted
 5 packet(s) received
 0.00% packet loss
 round-trip min/avg/max = 20/20/20 ms
[RTC]ping 1.1.1.1
 PING 1.1.1.1: 56  data bytes, press CTRL_C to break
Reply from 1.1.1.1: bytes=56 Sequence=1 ttl=254 time=50 ms
Reply from 1.1.1.1: bytes=56 Sequence=2 ttl=254 time=40 ms
Reply from 1.1.1.1: bytes=56 Sequence=3 ttl=254 time=30 ms
Reply from 1.1.1.1: bytes=56 Sequence=4 ttl=254 time=40 ms
Reply from 1.1.1.1: bytes=56 Sequence=5 ttl=254 time=10 ms

--- 1.1.1.1 ping statistics ---
 5 packet(s) transmitted
 5 packet(s) received
 0.00% packet loss
 round-trip min/avg/max = 10/34/50 ms
[RTC]ping 3.3.3.3
PING 3.3.3.3: 56  data bytes, press CTRL_C to break
Reply from 3.3.3.3: bytes=56 Sequence=1 ttl=255 time=30 ms
Reply from 3.3.3.3: bytes=56 Sequence=2 ttl=255 time=20 ms
Reply from 3.3.3.3: bytes=56 Sequence=3 ttl=255 time=40 ms
 Reply from 3.3.3.3: bytes=56 Sequence=4 ttl=255 time=30 ms
 Reply from 3.3.3.3: bytes=56 Sequence=5 ttl=255 time=20 ms
```

```
--- 3.3.3.3 ping statistics ---
5 packet(s) transmitted
5 packet(s) received
0.00% packet loss
 round-trip min/avg/max = 20/28/40 ms
```
#提示：通过ping命令反馈信息可知，路由器间所有网络都是互通的，各部门用户网关设置在路由器上，如果用户到路由器的二层网络没有问题，则各个部门间可以通过路由器实现三层互通

小贴士

（1）小型网络中的路由器比较少，部署OSPF时可以直接使用单区域，但是如果部署的是一个大中型企业网络，设备数量比较多，那么无论是从管理效率还是路由学习效率考虑，都需要部署多区域OSPF。

OSPF（多区域）

（2）OSPF路由域内的路由器需要通过router id进行唯一标识，就像每位员工的工号或每位学生的学号不能冲突一样。在一般情况下，可优先选用路由器上已经存在Loopback接口的IP地址作为OSPF路由器的router id。

（3）当利用network命令发布路由器接口所在网络到OSPF区域时，使用的是该接口IP地址的反掩码。

任务总结与思考

本任务重点讲述了在路由器上如何通过动态路由协议OSPF（单区域）实现路由信息的互相学习和传递，并且介绍了路由器接口的OSPF认证的配置。在现实中，如果遇到简单的小型网络，就可以考虑用OSPF（单区域）来实现配置。

思考以下两个问题。

1. 如果一台OSPF路由器的router id已经设置成功，但是配置有误，与规划的router id不一致，则应如何处理？若直接通过命令修改该id，能否立即启用新id？

2. 如果RTC的Loopback0接口被误删，那么网络中的OSPF路由会发生什么变化？

知识补给

4.2.1　OSPF的报文

OSPF启动之后，路由器与路由器之间可以通过OSPF报文来交互链路状态信息（协议的报文就像该协议的语言）。OSPF直接运行在IP上，使用的协议号为89，OSPF共有以下5种报文。

1. Hello报文：最常见的一种报文，用于发现、维护邻居关系，并在广播和NBMA类型的网络中选举DR和BDR。

2. DD报文：当两台路由器进行LSDB数据库同步时，用来描述发送端本身的LSDB。

3. LSR 报文：当路由器利用其他 OSPF 路由器发来的数据描述信息查漏之后，它就会发送链路状态请求消息，请求该设备用其链路状态数据库中的 LSA 来为发送端本身补缺。

4. LSU 报文：当接收到邻居发送过来的链路状态请求消息之后，路由器就会按照该消息中的请求，将邻居需要的 LSA 封装到 LSU 报文中发送给邻居。

5. LSACK 报文：当通过 LSU 接收到需要的 LSA 之后，路由器需要通过 LSACK 报文向对端设备进行确认。

4.2.2 OSPF 支持的网络类型

OSPF 支持的网络类型如表 4-2-2 所示。

表 4-2-2　OSPF 支持的网络类型

网络类型	常用的数据链路层封装	是否选举 DR/BDR
点到点（P2P）	PPP、HDLC	否
广播（Broadcast）	以太网	是
非广播多路访问（NBMA）	帧中继、ATM、X.25	是
点到多点（P2MP）	无	否

任务拓展

1. OSPF 定义了 4 种类型的网络，其中两种需要选举 DR（指定路由器）和 BDR（备份指定路由器），属于广播型网络和非广播多路访问型网络，都属于多路访问的网络，即这类网络可以连接多台设备，而且只要连接到网络中的任意设备就可以通过二层设备连接的网络访问其他设备，具体拓扑图如图 4-2-2 所示。

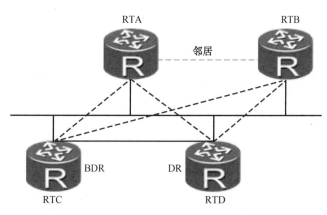

图 4-2-2　OSPF 网络中的 DR 和 BDR 拓扑图

2. 在上述的两种网络类型中，为了避免因任意路由器直接交互链路状态信息而带来的邻居关系太多和 LSA 的交互量太大等问题，引入了 DR 和 BDR 的概念，DR 充当从本网段的所有路由器收集链路状态信息，并将信息发布给其他路由器的专员角色。当 DR 出现故障时，

BDR 会取而代之。

小贴士

DR 和 BDR 是通过选举产生的，不是随机选择的，根据网络中连接的路由器的接口优先级进行 DR 选举。接口优先级的取值范围为 0~255，取值越大越优先，如果优先级相同，则比较设备的 router id，取值越大越优先。

小技巧

如图 4-2-3 所示，两台路由器 RTA 和 RTB 通过千兆以太网口互连，互联网为以太网，属于广播型网络，需要选举 DR 和 BDR，但是，在这种网络中，无论是否选举 DR，两台路由器都需要建立邻居关系交互 LSA，为了避免因它们默认选举 DR 和 BDR 而带来额外的网络开销，可以将接口的 OSPF 网络类型由默认的广播改为点到点，从而可以避免选举 DR 和 BDR。配置命令如下：

```
[RTA]interface GigabitEthernet 0/0/0
[RTA-GigabitEthernet0/0/0]ospf network-type ?
 broadcast  Specify OSPF broadcast network
 nbma       Specify OSPF NBMA network
 p2mp       Specify OSPF point-to-multipoint network
 p2p Specify OSPF point-to-point network
[RTA-GigabitEthernet0/0/0]ospf network-type p2p
[RTB]interface GigabitEthernet 0/0/0
[RTB-GigabitEthernet0/0/0]ospf network-type ?
 broadcast  Specify OSPF broadcast network
 nbma       Specify OSPF NBMA network
 p2mp       Specify OSPF point-to-multipoint network
 p2p Specify OSPF point-to-point network
[RTB-GigabitEthernet0/0/0]ospf network-type p2p
```

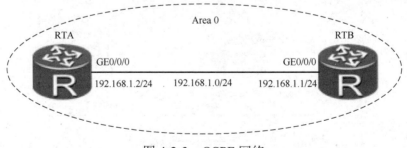

图 4-2-3 OSPF 网络

4.3 转换网络地址

➢ 任务情景

宇信公司采用联通公司的互联网接入服务，由于 IPv4 地址紧缺，联通公司只分配了 200.0.0.1~200.0.0.5 这 5 个地址用于互联网通信。宇信公司内部通信使用的是 192.168.1.0 网段地址。小李决定在公司的出口路由器上部署网络地址转换 NAT，以使私网用户可以访问公网。

任务分析

- 学会配置访问控制列表 ACL；
- 学会配置网络地址转换 NAT。

实施准备

1. HUAWEIAR2220E 路由器 1 台；
2. PC 2 台；
3. 交叉型 UTP 2 根；
4. Console 通信线缆 1 根。

实施步骤

1. 按要求连接网络设备，如图 4-3-1 所示。

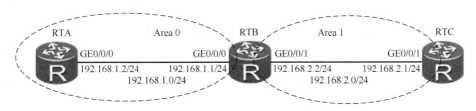

图 4-3-1　NAT 配置拓扑图

2. 按任务要求规划地址、子网掩码和网关，设备地址分配如表 4-3-1 所示。PC 和服务器地址配置分别如图 4-3-2 和图 4-3-3 所示。

表 4-3-1　设备地址分配表

设备名称	接口	IP 地址	子网掩码（Mask）	网关
RTA	GE 0/0/0	192.168.1.254	255.255.255.0	无
	GE 0/0/1	100.1.1.2	255.255.255.0	无
PC1	Ethernet	192.168.1.1	255.255.255.0	192.168.1.254
HTTP 服务器	Ethernet	100.1.1.1	255.255.255.0	100.1.1.2
地址池		200.0.0.1～200.0.0.5		

图 4-3-2　PC 地址配置

图 4-3-3　服务器地址配置

3. 配置 RTA 的设备名称及接口地址。

```
<Huawei>system-view   //进入系统视图
Enter system view, return user view with Ctrl+Z.
#提示：返回用户视图，请同时按<Ctrl>+<Z>键
[Huawei]sysname RTA    //路由器启动后，默认设备名称为Huawei，此处修改为RTA
[RTA]interface GigabitEthernet 0/0/0//进入GE0/0/1接口配置视图
[RTA-GigabitEthernet0/0/0]ip address 192.168.1.1 24   //配置GE0/0/1接口地址为192.168.1.1，并设置掩码为24位
#提示：24位掩码即为255.255.255.0
Dec 12 2018 16:32:24+00:00 RTA %%01IFNET/4/LINK_STATE(l)[0]:The line protocol IP on the interface GigabitEthernet0/0/0has entered the UP state.
#提示：此处系统提示接口GE0/0/1的协议状态为UP，说明接口配置成功
[RTA-GigabitEthernet0/0/0]quit   //退出GE0/0/0接口配置视图
[RTA]interface GigabitEthernet 0/0/1//进入GE0/0/1接口配置视图
[RTA-GigabitEthernet0/0/1]ip address 100.1.1.1 24   //配置GE0/0/1接口地址为100.1.1.1、掩码为255.255.255.0
Dec 12 2018 16:32:58+00:00 RTA %%01IFNET/4/LINK_STATE(l)[1]:The line protocol IP on the interface GigabitEthernet0/0/1 has entered the UP state.
#提示：此处系统提示接口GE0/0/1的协议状态为UP，说明接口配置成功
[RTA-GigabitEthernet0/0/1]quit   //返回系统视图
[RTA]   //返回系统视图
```

4. 在 RTA 上创建访问控制列表 ACL，匹配特定的流量并进行 NAT 转换，特定的流量为 PC1 发起的流量。

```
[RTA]acl 2000     // 创建基本的ACL，编号为2000
#提示：2000指ACL标号，属于可用范围
[RTA-acl-basic-2000]rule 5 permit source 192.168.1.0 0.0.0.255
//指定规则5允许来自192.168.1.0网段的流量
#提示：permit表示允许数据通过，若此处为Deny，则表示拒绝数据包通过；source表示数据包的来源地址，0.0.0.255表示反掩码
```

5. 在 RTA 上配置动态 NAT。

```
[RTA]nat address-group 1 200.0.0.1 200.0.0.5        //创建地址池1，并向地址池里放置一段地址
[RTA]interface GigabitEthernet 0/0/1     //进入GE0/0/1接口配置视图
[RTA-GigabitEthernet0/0/1]nat outbound 2000 address-group 1 no pat
//将ACL与地址池1关联起来，使得匹配ACL 2000的数据报文的源地址选用地址池中的某个地址，并进行NAT转换
```

6. 在 RTA 上查看地址池。

```
[RTA]dis nat address-group 1      //查看地址池1

NAT Address-Group Information:
-----------------------------------
Index   Start-address     End-address
-----------------------------------
1         200.0.0.1         200.0.0.5
-----------------------------------
Total : 1
#提示：地址池的起止地址分别是200.0.0.1与200.0.0.5，与表4-3-1规划一致
```

7. 检测 PC1 到 PC2 的连通性。

```
C:\Users\ >ping 100.0.0.1
```

```
正在 Ping 100.0.0.1 具有 32 字节的数据:
来自 100.0.0.1 的回复: 字节=32 时间=1ms TTL=64
来自 100.0.0.1 的回复: 字节=32 时间=1ms TTL=64
来自 100.0.0.1 的回复: 字节=32 时间=1ms TTL=64
来自 100.0.0.1 的回复: 字节=32 时间=1ms TTL=64

100.0.0.1 的 Ping 统计信息:
数据包: 已发送 = 4, 已接收 = 4, 丢失 = 0 (0% 丢失),
往返行程的估计时间(以毫秒为单位):
最短 = 1ms, 最长 = 1ms, 平均 = 1ms
#提示: PC1通过ping命令向IP地址为100.1.1.2的PC2发送4个ICMP包, 收到了4个回复, 每个包的字节数都为
32, 生存时间为64, 耗费时间为1ms, 丢包率为0%, 说明二者之间的网络互通
```

任务总结与思考

本任务重点讲述了在某单位网络的出口路由器上部署网络地址转换 NAT，使得网络中的用户可以访问互联网，同时介绍了访问控制列表 ACL 的方法。

思考以下两个问题。

1. 为什么把 IP 地址分为互联网（公网）地址和内部网（私网）地址？不进行地址转换的内网用户可以访问互联网吗？

2. 当 ACL 用来限制流量时，可以指定特定源地址和目的地址吗？用哪类 ACL 可以控制流量的源地址和目的地址呢？

知识补给

NAT 配置

4.3.1 网络地址转换 NAT

公有地址：由 NIC（网络信息中心）或 ISP（网络服务提供商）分配的地址，是全球统一的可寻址的地址。

私有地址：也叫私网地址，属于非注册地址，专门提供给组织机构在内部使用。因特网分配编号委员会（IANA）保留了 3 块地址作为私有地址：

10.0.0.0～10.255.255.255

172.16.0.0～172.16.255.255

192.168.0.0～192.168.255.255

随着网络设备的数量不断增长，用户对 IPv4 地址的需求也不断增加，导致可用的 IPv4 地址将逐渐耗尽。解决 IPv4 地址枯竭问题的权宜之计是向企业内部和家庭分配可重复使用的各类私网地址段。但是，私有地址不能在公网中路由，即私网主机不能与公网通信，也不能通过公网与其他私网通信。因此，私网用户若想访问互联网就必须在网络出口配置 NAT 转换。

NAT 转换方式有源 NAT 和 NAT Server 两种，适用于不同的场景。NAT 类型如表 4-3-2 所示。

表 4-3-2 NAT 类型

NAT 转换方式	类型	子类	含义	场景
源 NAT	NAT No-PAT	静态 NAT	只转换报文的 IP 地址，不转换接口	需要上网的私网用户数量较少，公网 IP 地址数量与同时上网的最大私网用户数量基本相同
		动态 NAT		
		Easy IP		
	NAPT		同时转换报文的 IP 地址和接口	公网 IP 地址数量少，需要上网的私网用户数量大
NAT Server	NAT Server		可根据具体场景来选择只转换报文的目的 IP 地址，还是同时转换目的 IP 地址和接口	私网服务器对公网用户提供服务

4.3.2 ACL 访问控制列表

ACL 分为基本 ACL、高级 ACL 和二层 ACL。

基本 ACL 可以使用报文的源 IP 地址、分片标记和时间段信息来匹配报文，其编号取值范围是 2000～2999。

高级 ACL 可以使用报文的源/目的 IP 地址、源/目的接口号及协议类型等信息来匹配报文，其编号取值范围是 3000～3999。

二层 ACL 可以使用源/目的 MAC 地址及二层协议类型等二层信息来匹配报文，其编号取值范围是 4000～4999。

任务拓展

1．NAT 转换的作用仅是将私网地址转换为公网地址吗？是否还有其他作用？
2．NAT 转换有缺点吗？如果想彻底解决 IPv4 地址短缺的问题，应该怎么做？

4.4 动态分配 IP 地址

动态分配 IP 地址

➢ 任务情景

宇信公司搬入新的办公园区后，员工们发现他们的计算机无法获取 IP 地址与网络参数，需要网络管理员小李给他们逐一配置 IP 地址，操作烦琐且不易管理。小李在高工程师的指导下，给网络部署了 DHCP 服务器，来动态地给每个人的终端分配 IP 地址。

主机 A 是宇信公司员工小王的计算机，她需要通过 DHCP 服务器自动获取 IP 地址、网关等参数。RTA 作为主机 A 的网关，小李决定将在 RTA 部署 DHCP 服务器。

➢ 任务分析

- ➢ 学会在路由器上配置 DHCP 服务；
- ➢ 掌握全局地址池的配置方法；
- ➢ 了解接口地址池；
- ➢ 学会在终端计算机上查看获取的地址等网络参数信息。

➢ 实施准备

1. HUAWEI AR2220E 路由器 1 台；
2. PC 1 台；
3. 交叉型 UTP 1 根；
4. Console 通信线缆 1 根。

➢ 实施步骤

1. 按要求连接网络设备，DHCP 配置拓扑图如图 4-4-1 所示。

图 4-4-1　DHCP 配置拓扑图

2. 按任务要求规划地址池参数信息、网关信息。地址池参数分配表如表 4-4-1 所示，主机 A 地址配置如图 4-4-2 所示。

表 4-4-1　RTA 地址池参数分配表

	地址池名称	Pool1
	地址池类型	全局地址池
参数名称	地址池分配的 IP 地址段	192.168.1.0
	保留的地址段	192.168.1.2～192.168.1.5
	租期	10 天
	网关	GE0/0/1 192.168.1.1/24

图 4-4-2 主机 A 地址配置

3. 配置 RTA 的设备名称及接口 IP 地址。

```
<Huawei>system-view   //进入系统视图
Enter system view, return user view with Ctrl+Z.
#提示：返回用户视图，请同时按<Ctrl>+<Z>键
[Huawei]sysname RTA   //路由器启动后默认设备名称为Huawei,此处修改为RTA
[RTA]interface GigabitEthernet 0/0/1/进入GE0/0/1接口配置视图
[RTA-GigabitEthernet0/0/1]ip address 192.168.1.1 24   //配置GE0/0/1接口地址为192.168.1.1，并设置掩码为24位
#提示：24位掩码即为255.255.255.0,此处IP地址的配置需与后面地址池里的网关配置保持一致
Dec 12 2018 16:32:24+00:00 RTA %%01IFNET/4/LINK_STATE(l)[0]:The line protocol IP on the interface GigabitEthernet0/0/1 has entered the UP state.
#提示：此处系统提示接口GE0/0/1的协议状态为UP,说明接口配置成功
[RTA-GigabitEthernet0/0/1]quit   //返回系统视图
```

4. 在 RTA 上启用 DHCP 服务。

```
[RTA]dhcp enable   //在全局视图中开启DHCP服务
Info: The operation may take a few seconds. Please wait for a moment.done.
//提示：这个操作将花费一点时间，请耐心等待
#提示：在默认情况下，DHCP功能不开启，需要在路由器上开启DHCP功能
```

5. 创建地址池并配置地址池参数信息。

```
[RTA]ip pool Pool1       //创建名称为Pool1的地址池
Info: It's successful to create an IP address pool.   //提示成功创建地址池
[RTA-ip-pool-pool1]network192.168.1.0 mask 24   //定义此地址池内的有效地址为192.168.1.0/24网段
#提示：地址池负责为终端设备分配IP地址，因此要将地址段先放入地址池内。如果我准备为同学们每人发一个苹果，这就需要先拿一个箱子把要发给大家的苹果装起来，而这里的IP地址相当于苹果，地址池就相当于放了很多苹果的箱子
[RTA-ip-pool-pool1]gateway-list 192.168.1.1       //设置网关地址
[RTA-ip-pool-pool1]excluded-ip-address 192.168.1.2 192.168.1.5
//保留的地址段为192.168.1.2~192.168.1.5
#提示：有些地址是用作DNS(域名解析协议)服务器地址或有其他用途的，这些地址需要在地址池里特别标注出来，表示不会分配给终端设备使用
[RTA-ip-pool-pool1]lease day 10       //租期10天
```

#提示：DHCP服务器以租借的形式向终端设备分配IP地址，租期即租借的期限。在默认情况下，租期是1天
　　[RTA-ip-pool-pool1]quit　　　　//返回系统视图

6. 在接口下开启DHCP服务功能。

　　[RTA]interface GigabitEthernet0/0/1　　　　//进入GE0/0/1接口配置视图
　　[RTA-GigabitEthernet0/0/1]dhcp select global　　　　//开启DHCP功能并且指定地址池类型为全局地址池模式
　　#提示：地址池的配置有两种模式，即全局地址池和接口地址池，这里使用全局地址池模式

7. 在RTA上查看地址池Pool1的参数信息。

```
[RTA]display ip pool name pool1        //查看地址池Pool1的参数信息
Pool-name          : pool1
 Pool-No           : 0
Lease              : 10 Days 0 Hours 0 Minutes

...//省略部分与本任务无关的屏幕提示信息

Gateway-0          : 192.168.1.1
 Mask              : 255.255.255.0
 VPN instance      : --
-----------------------------------------------------------------
Start           End          Total  Used  Idle(Expired)  Conflict  Disable
-----------------------------------------------------------------
192.168.1.1   192.168.1.254   253    0    249(0)            0         4
-----------------------------------------------------------------
```
#提示：Start表示地址池的起始地址为192.168.1.1，End表示地址池的结束地址为192.168.1.254。由于在前面的配置中把192.168.1.2~192.168.1.5保留了，所以此处的Disable，即禁用地址为4

8. 在终端设备（主机A）上检测是否获取到IP地址。

在Windows系统中打开命令提示符，如图4-4-3所示。

图4-4-3　打开命令提示符

输入如下命令，以检测是否获取到IP地址。

```
C:\Users\ >ipconfig    //在命令提示符里输入ipconfig命令，以查看IP地址等参数的获取情况
...//省略部分与本任务无关的屏幕提示信息
以太网适配器 以太网：
连接特定的 DNS 后缀 . . . . . . . . . :
IPv4 地址 . . . . . . . . . . . . : 192.168.1.254
子网掩码  . . . . . . . . . . . . : 255.255.255.0
默认网关. . . . . . . . . . . . . :192.168.1.1
...//省略部分与本任务无关的屏幕提示信息
```
#提示：此时主机A已经获取到IP地址192.168.1.254，网关地址为192.168.1.1，说明DHCP服务器部署成功

任务总结与思考

本任务重点讲述了路由器上配置 DHCP 服务的方法，同时介绍了如何在终端设备上查看自动获取的 IP 地址等网络参数。

思考以下两个问题。

1. 如果与终端设备相连接的接口的 IP 地址和地址池里指定的网关参数配置不一致，那么终端设备还能够获取 IP 地址等参数吗？为什么？

2. 配置地址池时，为什么要保留部分参数不分配？

 知识补给

ARG3 系列路由器和 X7 系列交换机等华为主流路由器和交换机都可以作为 DHCP 服务器，为主机及其他网络设备分配地址与网络参数。DHCP 服务器的地址池用来定义分配给主机的地址范围，分为接口地址池和全局地址池两种。

1. 接口地址池为连接到同一网段的主机或终端分配地址。地址池不需要创建，在为相连的终端设备分配地址时，默认使用与终端设备相连接的接口地址所在的网段。可以在服务器的接口执行 dhcp select interface 命令，使 DHCP 服务器采用接口地址池为客户端分配地址。

2. 全局地址池为所有连接到 DHCP 服务器的终端分配地址。可以在服务器的接口执行 dhcp select global 命令，使 DHCP 服务器采用全局地址池为客户端分配地址。

接口地址池的优先级比全局地址池高。配置全局地址池后，如果又配置了接口地址池，则客户端将优先从接口地址池中获取地址。

任务拓展

本任务在场景不变的情况下，将如图 4-4-1 所示的拓扑图中的路由器 RTA 换为一台三层交换机，尝试在该三层交换机上配置并使用 DHCP 服务，并实现为终端分配地址的功能。

小技巧

在命令提示符下，输入 ipconfig/release 命令，可以将已经获取的地址释放并断开网络连接；输入 ipconfig/renew 命令，可以重新获取 IP 地址并接入网络。

4.5 分发静态路由协议

静态路由重分发

➢ 任务情景

宇信公司需要在内部网络中部署 OSPF，以达到内网相互访问的目的。因业务发展需要，财务部小王希望访问 B 公司的一台 Web 服务器。小李决定在公司网络出口路由器 RTA 上部署静态路由，并将静态路由引入 OSPF 的内部网络。

➢ 任务分析

- 掌握 OSPF 的基础配置；
- 掌握在 OSPF 网络中重分发静态路由。

➢ 实施准备

1. HUAWEI AR2220E 路由器 3 台；
2. 交叉型 UTP 4 根；
3. Console 通信线缆 1 根；
4. PC 2 台。

➢ 实施步骤

1. 按要求连接网络设备，分发静态路由拓扑图如图 4-5-1 所示。

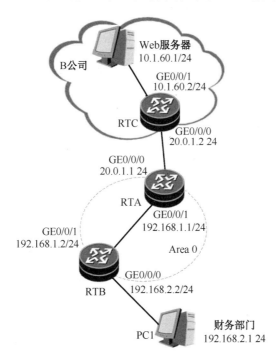

图 4-5-1　分发静态路由拓扑图

2. 按任务要求规划地址、子网掩码和网关，设备地址分配表如表 4-5-1 所示。PC 及 Web 服务器地址配置分别如图 4-5-2 和图 4-5-3 所示。

表 4-5-1 设备地址分配表

设备名称	接口	IP 地址	子网掩码（Mask）	网关
RTA	GE0/0/0	20.0.1.1	255.255.255.0	无
	GE0/0/1	192.168.1.1	255.255.255.0	无
RTB	GE0/0/0	192.168.2.2	255.255.255.0	无
	GE0/0/1	192.168.1.2	255.255.255.0	无
RTC	GE0/0/0	20.0.1.2	255.255.255.0	无
	GE0/0/1	10.1.60.2	255.255.255.0	无
PC1	Ethernet	192.168.2.1	255.255.255.0	192.168.2.2
Web 服务器	Ethernet	10.1.60.1	255.255.255.0	10.1.60.2

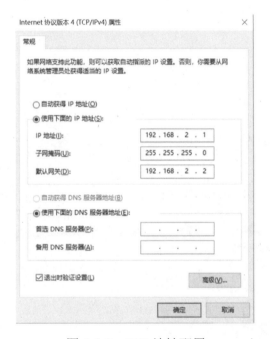

图 4-5-2 PC1 地址配置

图 4-5-3 Web 服务器地址配置

3. 配置 RTA 设备名称及接口 IP 地址。

```
<Huawei>system-view   //进入系统视图
Enter system view, return user view with Ctrl+Z.
#提示：返回用户视图时，请同时按<Ctrl>+<Z>键
[Huawei]sysname RTA    //路由器启动后默认设备名称为Huawei,此处修改为RTA
[RTA]interface GigabitEthernet 0/0/0//进入GE0/0/0接口配置视图
  [RTA-GigabitEthernet0/0/0]ip address 10.1.60.2 24    //配置GE0/0/0接口地址为10.1.60.2，并设置掩码为24位
  #提示：24位掩码为255.255.255.0
  Dec 12 2018 16:32:24+00:00 RTA %%01IFNET/4/LINK_STATE(l)[0]:The line protocol IP on the interface GigabitEthernet0/0/0 has entered the UP state.
  #提示：此处系统提示接口GE0/0/0的协议状态为UP，说明接口配置成功
  [RTA-GigabitEthernet0/0/0]quit   //返回接口GE0/0/0配置视图
```

[RTA]　//返回系统视图
　　[RTA]interface GigabitEthernet 0/0/01//进入GE0/0/1接口配置视图
　　[RTA-GigabitEthernet0/0/1]ip address 192.168.1.1 24　　//配置GE0/0/1接口地址为192.168.1.1，并设置掩码为24位
　　#提示：24位掩码为255.255.255.0
　　Dec 12 2018 16:32:24+00:00 RTA %%01IFNET/4/LINK_STATE(l)[0]:The line protocol IP on the interface GigabitEthernet0/0/1 has entered the UP state.
　　#提示：此处系统提示接口GE0/0/1的协议状态为UP，说明接口配置成功
　　[RTA-GigabitEthernet0/0/1]quit　　//返回接口GE0/0/0配置视图
　　[RTA]　//返回系统视图

4. 配置RTB设备名称及接口地址。

　　<Huawei>system-view　//进入系统视图
　　Enter system view, return user view with Ctrl+Z.
　　#提示：返回用户视图，请同时按下<Ctrl>+<Z>键
　　[Huawei]sysname RTB　　//路由器启动后默认设备名称为Huawei,此处修改为RTB
　　[RTB]interface GigabitEthernet 0/0/0//进入GE0/0/0接口配置视图
　　[RTB-GigabitEthernet0/0/0]ip address 192.168.2.2 24　　//配置GE0/0/0接口地址为192.168.2.2，并设置掩码为24位
　　#提示：24位掩码为255.255.255.0
　　Dec 12 2018 16:32:24+00:00 RTB %%01IFNET/4/LINK_STATE(l)[0]:The line protocol IP on the interface GigabitEthernet0/0/0 has entered the UP state.
　　#提示：系统提示接口GE0/0/0的协议状态为UP，说明接口配置成功
　　[RTB-GigabitEthernet0/0/0]quit　　//返回接口GE0/0/0配置视图
　　[RTB]　//返回系统视图
　　[RTB]interface GigabitEthernet 0/0/01//进入GE0/0/1接口配置视图
　　[RTB-GigabitEthernet0/0/1]ip address 192.168.1.2 24　　//配置GE0/0/1接口地址为192.168.1.2，并设置掩码为24位
　　#提示：24位掩码为255.255.255.0
　　Dec 12 2018 16:32:24+00:00 RTB %%01IFNET/4/LINK_STATE(l)[0]:The line protocol IP on the interface GigabitEthernet0/0/1 has entered the UP state.
　　#提示：此处系统提示接口GE0/0/1的协议状态为UP，说明接口配置成功
　　[RTB-GigabitEthernet0/0/1]quit　　//返回接口GE0/0/0配置视图
　　[RTB]　//返回系统视图

5. 配置RTC设备名称及接口地址。

　　<Huawei>system-view　//进入系统视图
　　Enter system view, return user view with Ctrl+Z.
　　#提示：返回用户视图，请同时按<Ctrl>+<Z>键
　　[Huawei]sysname RTC　　//路由器启动后默认设备名称为Huawei,此处修改为RTC
　　[RTC]interface GigabitEthernet 0/0/0//进入GE0/0/0接口配置视图
　　[RTC-GigabitEthernet0/0/0]ip address 20.0.1.2 24　　//配置GE0/0/0接口地址为20.0.1.2,并设置掩码为24位
　　#提示：24位掩码为255.255.255.0
　　Dec 12 2018 16:32:24+00:00 RTC %%01IFNET/4/LINK_STATE(l)[0]:The line protocol IP on the interface GigabitEthernet0/0/0 has entered the UP state.
　　#提示：系统提示接口GE0/0/0的协议状态为UP，说明接口配置成功
　　[RTC-GigabitEthernet0/0/0]quit　　//返回接口GE0/0/0配置视图
　　[RTC]　//返回系统视图
　　[RTC]interface GigabitEthernet 0/0/1//进入GE0/0/1接口配置视图
　　[RTC-GigabitEthernet0/0/1]ip address 10.1.60.2 24　　//配置GE0/0/1接口地址为10.1.60.2,并设置掩码为24位
　　#提示：24位掩码为255.255.255.0

```
Dec 12 2018 16:32:24+00:00 RTC %%01IFNET/4/LINK_STATE(l)[0]:The line protocol IP on
the interface GigabitEthernet0/0/1 has entered the UP state.
```
#提示：系统提示接口GE0/0/1的协议状态为UP，说明接口配置成功

```
[RTC-GigabitEthernet0/0/1]quit     //返回接口GE0/0/0配置视图
[RTC]                              //返回系统视图
```

6. 在 RTA 上配置 OSPF。

```
[RTA]ospf router-id 1.1.1.1        //启动OSPF并配置Router ID 为1.1.1.1
[RTA-ospf-1]area 0                 //划分区域area 0
[RTA-ospf-1-area-0.0.0.0]network 192.168.1.0 0.0.0.255
//将192.168.1.0网络宣告到OSPF网络中
```
#提示：OSPF宣告网络时，需要加反掩码

7. 在 RTB 上配置 OSPF。

```
[RTB]ospf router-id 2.2.2.2        //启动OSPF并配置Router ID 为2.2.2.2
[RTB-ospf-1]area 0                 //划分区域area 0
[RTB-ospf-1-area-0.0.0.0]network 192.168.1.0 0.0.0.255
//将192.168.1.0网络宣告到OSPF网络中
```
#提示：OSPF宣告网络时，需要加反掩码
```
[RTB-ospf-1-area-0.0.0.0]network 192.168.2.0 0.0.0.255
//将192.168.2.0网络宣告到OSPF网络中
```
#提示：OSPF宣告网络时，需要加反掩码

8. 查看 RTA 的路由表，此时路由表中已经学习到了一条 OSPF 路由。

```
[RTA]dis ip routing-table
Route Flags: R - relay, D - download to fib
------------------------------------------------------------------
Routing Tables: Public
        Destinations : 7       Routes : 7

Destination/Mask    Proto    Pre  Cost   Flags NextHop       Interface

20.0.1.0/24         Direct   0    0      D     20.0.1.1      igabitEthernet0/0/0
20.0.1.1/32         Direct   0    0      D     127.0.0.1     GigabitEthernet0/0/0
127.0.0.0/8         Direct   0    0      D     127.0.0.1     InLoopBack0
127.0.0.1/32        Direct   0    0      D     127.0.0.1     InLoopBack0
192.168.1.0/24      Direct   0    0      D     192.168.1.1   GigabitEthernet0/0/1
192.168.1.1/32      Direct   0    0      D     127.0.0.1     GigabitEthernet0/0/1
192.168.2.0/24      OSPF     10   2      D     192.168.1.2   GigabitEthernet0/0/1
```
#提示：注意此时RTA的路由表中出现了192.168.2.0/24的路由，协议（Proto）类型为OSPF，下一跳（FlagsNextHop）地址为192.168.1.2，即路由器的输出接口为GE0/0/1，说明RTA可以正常转发去往这个网络的数据包

9. 在 RTA 上配置静态路由。

```
[RTA]ip route-static 10.1.60.0 24 20.0.1.2    //配置去往10.1.60.0/24的路由
```
#提示：下一跳地址为20.0.1.2

10. 查看 RTA 的路由表。

```
[RTA]dis ip routing-table
Route Flags: R - relay, D - download to fib
------------------------------------------------------------------
Routing Tables: Public
        Destinations : 8       Routes : 8

Destination/Mask   Proto  Pre  Cost   Flags  NextHop      Interface
```

```
10.1.60.0/24      Static  60  0  RD  20.0.1.2     GigabitEthernet0/0/0
20.0.1.0/24       Direct  0   0  D   20.0.1.1     GigabitEthernet0/0/0
20.0.1.1/32       Direct  0   0  D   127.0.0.1    GigabitEthernet0/0/0
127.0.0.0/8       Direct  0   0  D   127.0.0.1    InLoopBack0
127.0.0.1/32      Direct  0   0  D   127.0.0.1    nLoopBack0
192.168.1.0/24    Direct  0   0  D   192.168.1.1
192.168.1.1/32    Direct  0   0  D   127.0.0.1    GigabitEthernet0/0/1
192.168.2.0/24    OSPF    10  2  D   192.168.1.2  GigabitEthernet0/0/1
```

#提示：注意此时RTA的路由表中出现了10.1.60.0/24的路由，协议（Proto）类型为STATIC，下一跳（FlagsNextHop）地址为20.0.1.2，即路由器的送出接口为GE0/0/0，说明RTA可以正常转发去往这个网络的数据包

11. 在RTA上配置静态路由重分发，将静态路由引入OSPF。

```
[RTA]ospf             //进入OSPF视图
[RTA-ospf-1]import-route static   //将静态路由重发布至OSPF网络
```

12. 查看RTB的路由表。

```
[RTB]display ip routing-table
Route Flags: R - relay, D - download to fib
------------------------------------------------------------
Routing Tables: Public
         Destinations : 7       Routes : 7

Destination/Mask   Proto   Pre  Cost  Flags  NextHop       Interface

10.1.60.0/24       O_ASE   150  1     D      192.168.1.1   GigabitEthernet0/0/1
127.0.0.0/8        Direct  0    0     D      127.0.0.1     InLoopBack0
127.0.0.1/32       Direct  0    0     D      127.0.0.1     InLoopBack0
192.168.1.0/24     Direct  0    0     D      192.168.1.2   GigabitEthernet0/0/1
192.168.1.2/32     Direct  0    0     D      127.0.0.1     GigabitEthernet0/0/1
192.168.2.0/24     Direct  0    0     D      192.168.2.2   GigabitEthernet0/0/0
192.168.2.2/32     Direct  0    0     D      127.0.0.1     GigabitEthernet0/0/0
```

#提示：此时RTB的路由表中出现了10.1.60.0/24的路由，协议（Proto）类型为O_ASE类型（OSPF的外部路由），下一跳（FlagsNextHop）地址为192.168.1.1，即路由器的送出接口为GE0/0/1，说明RTA可以正常转发去往这个网络的数据包

13. 在RTC上配置默认路由。

```
[RTC]ip route-static 0.0.0.0 0 20.0.1.1     //配置可以去往任意网络的默认路由
```
#提示：下一跳地址为20.0.1.1

14. 检测财务部小王的PC1与Web服务器之间的连通性。

```
C:\Users\ >ping 10.1.60.1

正在 Ping 10.1.60.1 具有 32 字节的数据:
来自 10.1.60.1 的回复: 字节=32 时间<1ms TTL=64
来自 10.1.60.1 的回复: 字节=32 时间=1ms TTL=64
来自 10.1.60.1 的回复: 字节=32 时间=1ms TTL=64
来自 10.1.60.1 的回复: 字节=32 时间<1ms TTL=64

10.1.60.1 的 Ping 统计信息:
    数据包: 已发送 = 4，已接收 = 4，丢失 = 0 (0% 丢失)，
    往返行程的估计时间(以毫秒为单位):
```

最短 = 0ms，最长 = 1ms，平均 = 0ms
#提示：PC1通过ping命令向IP为10.1.60.1的Web服务器发送4个ICMP包，收到了4个回复报文，每个包的字节数都为32，生存时间为64，耗费时间为1ms，丢包率为0%，说明二者之间的网络已经互通，状态良好

任务总结与思考

本任务重点讲述了如何将静态路由重分发至 OSPF 网络，同时展示了默认路由的配置方法。

思考以下两个问题。

1. 可以在 OSPF 网络中重分发其他动态路由协议吗？
2. 路由重分发解决了网络协议工作过程中的什么问题？

知识补给

路由重分发可以实现多种路由协议之间共享路由信息，并进行路由信息交换，使同一产生环路网络可以高效地支持多种路由协议。

路由重分发并不是完美的，它有如下几个缺点。

1. 产生环路：根据重分发的使用方法，路由器有可能将从某个 AS 收到的路由信息发回该 AS 中，这可能会产生环路。
2. 路由信息不兼容：不同的路由协议使用不同的度量值，因为这些度量值可能无法正确引入不同的路由协议，所以使用重分发的路由信息进行路径选择时，其结果可能不是最优的路径。
3. 收敛效率不一致：不同的路由协议收敛效率不同，如 RIP 比 OSPF 收敛慢，因此如果一条链路断掉，则 OSPF 网络将比 RIP 网络更早得知这一信息。

任务拓展

在如图 4-5-1 所示的场景中，如果 RTC 不允许配置默认路由，那么如何配置才能实现 PC1 访问 Web 服务器？

思考与实训 4

一、判断题

1. OSPF 可以支持多区域，骨干区域为 Area1。（   ）

2．OSPF 的路由聚合在任意路由器上都可以执行成功。（　　）

3．OSPF 路由聚合之后，明细路由和聚合路由都会传递出去，除非通过命令抑制明细路由。（　　）

4．路由聚合造成路由器学习不到明细路由，因此网络会不连通。（　　）

5．聚合的网段只要包含明细路由即可，无须考虑其他因素。（　　）

6．VRRP 备份组中只能有两台路由器，不能有两台以上路由器。（　　）

7．在一个 VRRP 备份组中可以存在多台 MASTER 路由器，这主要是为了负载均衡。（　　）

8．VRRP 状态机有 MASTER、BACKUP 两种。（　　）

9．OSPF 在点到点链路中无须选举 DR 和 BDR。（　　）

10．OSPF 不是一种静态路由协议。（　　）

11．在默认情况下，对于两个通过点到点链路相连的接口，IP 地址较高的接口会被选举为 DR，另一个接口则会被选举为 BDR。（　　）

12．DHCP 服务器只能为终端设备动态地分配 IP 地址，不能分配 DNS 地址等其他参数。（　　）

13．在全局地址池模式下，网关地址不用配置。（　　）

14．一个 VLAN 可以映射到多个实例。（　　）

15．一个实例只能映射一个 VLAN。（　　）

16．DHCP 只能为 PC 分配 IP 地址，不能为其他设备分配 IP 地址。（　　）

17．VLAN1 默认已经存在，且无法被删除。（　　）

18．OSPF 优先级比 RIP 高。（　　）

二、选择题

1．在使用 NETWORK 命令把接口（IP 地址为 10.0.8.10/24）加入 OSPF 进程 100 的区域 0 时，以下命令正确的是（　　）。

　　A．[AR1-ospf-100]network 10.0.8.10　255.255.255.0

　　B．[AR1-ospf-100]network 10.0.8.0　0.0.0.255

　　C．[AR1-ospf-100-area-0.0.0.0]network 10.0.8.10　255.255.255.0

　　D．[AR1-ospf-100-area-0.0.0.0] network 10.0.8.0　0.0.0.255

2．OSPF 的报文类型有（　　）。（多选）

　　A．Hello

　　B．数据库描述（DD）

　　C．链路状态请求（LSR）

　　D．链路状态更新（LSU）

E．链路状态确认（LSACK）

三、填空题

1．中小型企业网络采用的是典型的_____、_____、_____三层架构。
2．OSPF 的协议优先级是_____。
3．中小型企业网络里采用的拓扑结构中，不提供冗余的是_____。
4．OSPF 的度量值计算方法是_____。
5．默认路由的目的网络是_____。
6．RIP 规定超过_____跳为路由不可达。
7．OSPF 开销计算公式是_____。
8．路由互相引入的命令是 _____。
9．DHCP 的全称是_____。
10．DHCP 地址池的两种模式分别是 _____、_____。
11．DHCP 默认租期是_____。
12．网络地址转换 NAT 是指将_____转换为_____。
13．基本 ACL 的编号范围是_____。

四、实训操作

1．按照如图 4-1 所示拓扑图及情景要求，通过在三层交换机上启用 OSPF 实现全网互访。

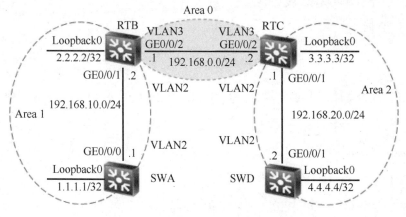

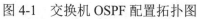

图 4-1　交换机 OSPF 配置拓扑图

交换机的 ospf

2．组网配置如图 4-2 所示，按照图中的区域规划接口规划配置，通过 OSPF（多区域）实现全网互通。

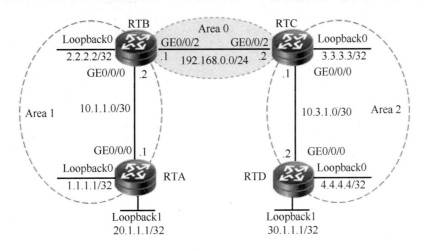

图 4-2　路由器 OSPF 配置拓扑图

项目 5

无线与安全设备

☆ 项目背景

宇信公司搬入新的企业园区后,部署了全新的企业网络,目前有线网络部分已实施完成,能够承载公司正常业务流量。为满足员工移动办公需求,以及向贵宾及访客提供便捷的无线接入服务,公司计划在会议室、开放办公区、楼道和洗手间等地方进行无线覆盖。在满足公司业务和办公需要的基础上,为保障企业内部网络信息安全,避免遭受来自互联网的攻击,公司购置了防火墙设备,代替路由器作为公司内部网络的出口设备,以保护网络数据安全。

5.1 认识无线设备

➢ **任务情景**

宇信公司的无线网络项目规划已经完成，确定使用技术成熟、市场保有率高的华为智能无线设备，主要包括 HUAWEIAC6005 无线控制器（AC）和 HUAWEIAP6050DN 无线接入点（AP）两类产品。在安装与部署智能无线设备之前，小李需要首先认识并熟悉产品的外观、性能及安装方法。

➢ **任务分析**

- 认识无线 AC 和无线 AP；
- 了解无线射频主要分类；
- 了解无线终端接入和漫游。

➢ **实施准备**

1. HUAWEIAC6005-8-PWR 无线接入控制器 1 台；
2. HUAWEIAP6050DN 1 台；
3. 笔记本电脑 1 台、智能手机 1 部；
4. 直通双绞线 2 根；
5. Console 通信线缆 1 根；
6. Secure CRT 仿真软件 1 套；
7. 防静电手套与裁纸刀。

➢ **实施步骤**

1. 打开一台 HUAWEIAC6005-8-PWR 的包装箱，观察该设备的外观结构（见图 5-1-1）。其中，序号 1 为 MODE 按钮，序号 7 为交流电源输入口，序号 4 为 Console 接口。

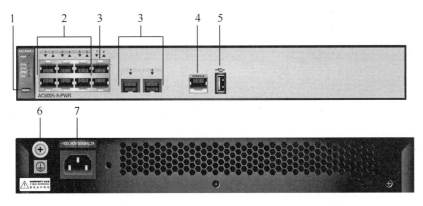

图 5-1-1　HUAWEIAC6005-8-PWR 外观结构示意图

2. 把 AC 接入机柜电源，用 Console 通信线缆连接 HUAWEIAC6005-8-PWR 与 PC，如图 5-1-2 所示。

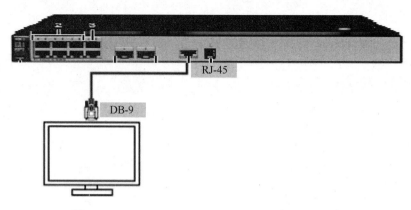

图 5-1-2　HUAWEIAC6005-8-PWR 与 PC 互连

3. 打开外部电源及电源模块上的开关，可以听到风扇旋转的声音，还可以看到通风孔处有空气排出。设备上的 PWR 灯和电源模块上的 STAT 灯绿色常亮，如图 5-1-3 所示。PWR 和 STAT 分别对应图中的序号 1 和序号 3。

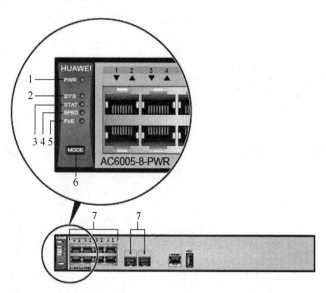

图 5-1-3　HUAWEIAC6005-8-PWR 指示灯

4. 打开 PC 上的 SecureCRT 软件，在"快速连接"对话框中设置传输速率为 9600bit/s、8 位数据位、1 位停止位，无奇偶校验和无流控，单击"连接"按钮，如图 5-1-4 所示。

5. 首次登录时，输入默认的用户名和密码后进入用户视图，并设置用户通过 Console 接口登录。

图 5-1-4　仿真终端软件参数设置

```
<AC6005>system-view                          //进入系统视图
[AC6005]user-interface console 0             //进入Console接口用户配置模式
[AC6005-ui-console0] user privilege level 15 //设置用户登录优先级为最高级别15
```

6. 配置 Console 用户界面的验证模式为密码验证。

```
[AC6005-ui-console0]authentication-mode password      //设置用户界面的验证模式为密码验证
[AC6005-ui-console0]set authentication password cipher //设置密码存储方式为cipher
Info: A plain text password is a string of 8 to 16 case-sensitive characters and must be a combination of at least two of the follow
      ing: uppercase letters A to Z, lowercase letters a to z, digits, and special characters (including spaces and the following :`~!@#$%
^&*()-_=+|[{}];:'",<.>/?). A cipher text password contains 56 or 68 characters.
#提示：密码长度为8～16个区分大小写的字母或数字符号，密码组成至少包括大/小写字符、数字、字符中的3种以上
Current Password: //提示：输入当前密码
New Password: //提示：输入新密码
Confirm New Password: //提示：再次确认新密码
[AC6005-ui-console0] quit
```

7. 验证配置结果。断开连接后重新登录，以检查新的密码是否生效。

```
[AC6005]user-interface console 0             //进入Console接口
[AC6005-ui-console0] display this            //查看Console接口信息
#
user-interface con 0
 authentication-mode password
 set authentication password cipher %^%#f6441rvzHS/8;5-QxO))c9JWSLakF@>eYf:[Ng6Y%^%#
#
return
#提示：Console用户界面已配置完成，用户可以通过Console接口使用Password方式登录设备，实现本地维护
```

8. 修改设备的主机名为 AC6005A。

```
<AC6005>system-view           //查看系统信息
[AC6005]sysname AC6005A       //修改设备名称
[AC6005A]
```

9. 修改系统时间和所在时区。

```
<AC6005A>clock datetime 10:10:00 2018-12-01    //修改系统时间为2018-12-01 10:10:00
<AC6005A> clock timezone BJ add 08:00:00       //修改系统时区为北京BJ
```
#提示：系统默认的UTC是伦敦时间，若伦敦当地时间为2018年12月1日0时0分0秒，则想要得到对应的北京时间的方法是：北京处于+8时区，时间偏移量增加了8。在配置时，需要在系统默认的UTC时区的基础上，加上偏移量8，就能得到预期的BJ时区

10. 连接 AP，打开 HUAWEIAP6050DN 包装箱，取出无线接入点 AP，用超 5 类双绞线缆把 AP 的 Ethernet 接口和 AC 的 GE0/0/1 接口连接起来，连接方式如图 5-1-5 所示，Ethernet 接口为图 5-1-6 中的接口 3。

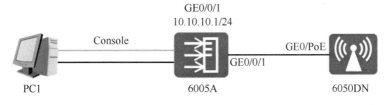

图 5-1-5 AP 连接方式示意

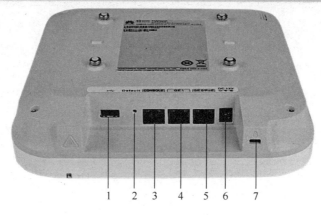

图 5-1-6　AP 外观

11. 查看并记录 AP 背板上的 MAC 地址及 S/N 编号，如图 5-1-7 所示。

图 5-1-7　AP 设备信息

12. 配置 AP 管理功能，在 AC 上配置源地址，便于 AP 上电后查找 AC。

```
[AC6005A]capwap source ip-address 10.10.10.1    //配置AC的源地址，便于AP查找AC
[AC6005A]dhcp enable
[AC6005A]ip pool ap-guanli              //创建地址池为分体AP和中心AP分配地址
[AC6005A -ip pool ap-guanli]network 10.10.10.0  mask 24  //地址池的范围为10.10.10.0，掩码为255.255.255.0
[AC6005A -ip pool ap-guanli]gateway-list 10.10.10.1    //地址池的网关为10.10.10.1
[AC6005A -ip pool ap-guanli]quit  //返回全局配置模式
[AC6005A]wlan
[AC6005A]ap-id 1 type-id 1  ap-mac7858601f9fe0 ap-sn 21500829352SJ2603550    //添加AP信息
```

13. 查看 AP 信息。

```
<AC6005A>display ap all
Total AP information:
nor  : normal           [1]
--------------------------------------------------------------------
ID   MAC         Name   Group    IP           Type        State STA Uptime
0    7858-601f-9fe0 area_1 ap-group1 10.10.10.254 AP6050DN-AGN  nor 0 4H:49M:11S
Total: 1
#提示：显示新添加的AP的信息，说明AP已与AC正常连接
```

 知识补给

1. HUAWEIAC6005-8-PWR 的接口及其功能说明，如图 5-1-8 及表 5-1-1 所示。

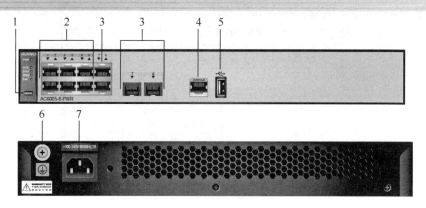

图 5-1-8　HUAWEIAC6005-8-PWR 外观

表 5-1-1　HUAWEIAC6005-8-PWR 接口及其功能说明

序号	接口名称	功能说明
1	MODE 按钮	模式切换按钮
2	6 个 10/100/1000BASE-T 以太网电接口	支持 10M/100M/1000M 自适应，支持 MDIX。HUAWEIAC6005-8-PWR 支持 6 个接口 PoE 供电
3	2 对 Combo 接口	支持 10M/100M/1000M 自适应，支持 PoE 供电
4	Console 接口	带外管理接口，支持 RJ45 连接器与 PC 的 RS232 连接
5	USB 接口	配合 U 盘使用，可用于传输配置文件和升级文件等
6	接地点	
7	交流电源输入接口	

2．HUAWEIAC6005-8-PWR 指示灯说明，如图 5-1-9 及表 5-1-2 所示。

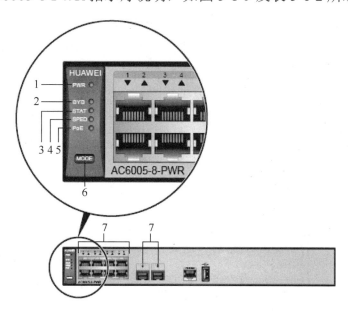

图 5-1-9　HUAWEIAC6005-8-PWR 指示灯

表 5-1-2 HUAWEIAC6005-8-PWR 指示灯说明

序号	指示灯名称	颜色	状态及含义说明
1	PWR（电源指示灯）	绿色	常亮：设备电源供电正常
		黄色	PoE 电源故障
		常灭	设备未上电
2	SYS（系统指示灯）	绿色	慢闪：系统正常运行中
			快闪：系统正在启动过程中
		红色	设备不能正常启动，或者运行中有温度、风扇异常报警
		常灭	系统未运行
3	STAT（模式状态）	绿色	常亮：业务接口指示灯为默认模式，即为 STAT 模式
		常灭	没有选择 STAT 模式
4	SPEED（模式状态灯）	绿色	常亮：表示业务接口指示灯暂时用来指示接口的速度，45 秒后恢复默认模式（STAT）
		常灭	表示没有选择 SPEED 模式
5	PoE（PoE 模式状态灯）	绿色	常亮：表示业务接口指示灯暂时用来指示各接口的 PoE 状态，45 秒后自动恢复默认模式（STAT）
		常灭	表示没有选择 PoE 模式
6	MODE（模式切换按钮）	红绿双色	按一次则 SPEED 灯亮绿色，此时业务接口指示灯暂时用来指示各接口的速率状态
			再按一次则 PoE 灯亮绿色，此时业务接口指示灯暂时用来指示各接口的 PoE 状态
			再按一次则恢复默认状态，即 STAT 灯亮绿色
7	GE 电接口指示灯	接口从 1 开始编号，顺序为从下到上、从左到右	SATA、SPEED、PoE 灯均绿色常亮：闪烁表示接口正在发送或接收数据，接口工作在 10M/100M 速率
			SATA、SPEED、PoE 灯均灭：表示接口无连接或被关闭，或未供电
	GE 光接口指示灯	参考三角形状的箭头指向	SPEED 灯绿色闪烁：表示此接口工作速率为 1000M
			PoE 灯黄色闪烁：表示供电 PoE 功能未使用或错误

3. HUAWEIAP6050DN 接口及其功能说明，如图 5-1-10 及表 5-1-3 所示。

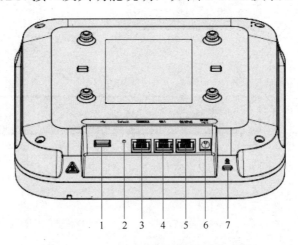

图 5-1-10 HUAWEIAP6050DN 外观

表 5-1-3　HUAWEIAP6050DN 接口及其功能说明

序号	接口名称	功能说明
1	USB	连接 U 盘或其他存储设备，用于扩展存储。支持 USB2.0 标准
2	Default	复位按钮，长按超过 3 秒恢复出厂默认值并重新启动
3	CONSOLE	控制口，连接维护终端，用于设备配置和管理
4	GE1	10Mbit/s /100Mbit/s /1000Mbit/s，用于有线以太网连接
5	GE0/PoE	10Mbit/s /100Mbit/s /1000Mbit/s，用于有线以太网连接，支持 PoE 输入
6	DC 12V	直流电源接口，用于连接 12V 电源适配器
7	防盗锁孔	连接防盗锁
8	天线接口	连接外置天线，用于发送和接收无线信号。接口类型为反极性 SMA 母头（RP-SMA-K）。仅适用于支持外置天线的 AP 款型

4. HUAWEIAP6050DN 指示灯及其说明，如图 5-1-11 及表 5-1-4 所示。

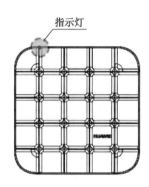

图 5-1-11　HUAWEIAP6050DN 指示灯

表 5-1-4　HUAWEIAP6050DN 指示灯说明

信息类型	颜色	指示灯状态	功能说明
上电默认状态	绿色	常亮	设备刚上电，软件未启动
软件启动过程状态	绿色	闪烁一下后常亮	软件启动过程状态
运行	绿色	慢闪（0.5Hz）	系统运行正常，以太网连接正常，表示有用户关联
运行	绿色	慢闪（0.2Hz）	系统运行正常，以太网连接正常，无用户关联，系统处于低功耗状态
报警	绿色	快闪（4Hz）	软件升级过程中。软件加载并启动后，在 FIT AP 或云管理工作模式下，进入请求上线状态。上线成功之前一直处于此状态。在 FIT AP 或云管理工作模式下，上线失败
故障	红色	常亮	设备有影响业务且无法自行恢复的故障（如 DRAM 检测失败、软件系统加载失败等），需人工干预

小贴士

大多数华为 AP 可以在胖模式和瘦模式这两种模式下工作，能够独立工作的 AP 称为 FAT AP（胖模式），本任务中与无线 AC 配合使用的 AP 称为 FIT AP（瘦模式），在这种工作模式下，配置都在无线 AC 上实施并下发到 AP，AP 零配置上线。

任务拓展

当网络规模较小时，一般采用直连式组网。本任务中采用的组网方式是直连式组网，在 AC 下直接接入 AP 或交换机，AC 同时扮演 AC 和汇聚交换机的角色，AP 的数据业务和管理业务都由 AC 集中转发和处理。VLAN 网络中通常有 AC 控制器、AP 接入点及交换机等设备，如图 5-1-12 所示。

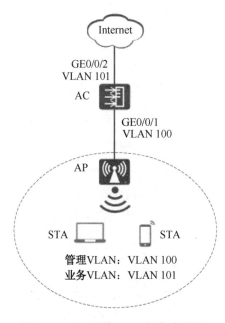

图 5-1-12　无线 AC 直连式组网

小贴士

本任务中部署的无线应用在公司室内，所以选用的是室内型号的 AP，如果是在广场上部署，则选用室外型号的 AP，同时需要考虑 AP 能够承受的温度范围，以及防尘、防水、防雷等因素。

小贴士

同一个局域网内有多个 AP 接入时，发射 2.4G 或 5G 射频信号所用的信道应当避开干扰较多的信道，使用户有较好的网络体验。

小技巧

无线控制器 HUAWEIAC6005 系列设备支持 NAT、DHCP 特性，相关配置与路由器或三层交换机类似，如果企业规模较小，则可以使用 AC6005 系列设备作为出口设备与广域网互连。当无线控制器接入的 AP 数量较多时（默认授权为 8 个），则需要购买 AP 接入授权，才能管理更多的 AP。

5.2 规划 WLAN

➤ 任务情景

宇信公司办公区共有 5 层，员工 200 人，无线接入需求主要集中在市场部、会议室、餐厅、来宾休息区等地方，因此要求在这些区域内能够实现移动办公、无线漫游等功能。高工程师带领小李做了实地勘测，并进行了 WLAN 规划。

➤ 任务分析

- 获取网络部署的必要信息（包括客户预算）；
- 确认无线覆盖环境中存在的障碍物及信号通过障碍物的衰减程度；
- 确定 AP 的选型和数量；
- 确定 AP 的安装位置和供电方式。

➤ 实施准备

1. 笔记本电脑 1 台；
2. 智能手机 1 部；
3. 华为 WLAN Planner 软件；
4. 测距工具。

➤ 实施步骤

1. 需求分析。

高工程师带领小李与宇信公司业务负责人王经理对接后了解到，宇信公司的主要业务内容为国内农产品出口贸易，公司员工总数为 200 人，无线业务主要集中在市场部、会议室、餐厅、来宾休息区等地方，使用人数大约为 40 人，要求每个用户使用带宽为 2Mbp，整个公司业务并发率约为 75%。

【提示：并发率指使用无线网络的用户占总连接无线网络的用户的比例。例如，有些用户保持无线连接，但并没有上传或下载数据。】

2. 现场工勘。

宇信公司的无线覆盖环境包括办公区、会议室及客户休息区等，如图 5-2-1 所示。高工程师认为，公司的无线覆盖环境处于室内，属于半开放环境，有玻璃、石膏隔断和承重柱等障碍物。

图 5-2-1 宇信公司无线覆盖环境

3．方案设计。

根据用户需求分析和现场工勘结果，按照无线技术及 WLAN802.11MAC 架构，高工程师给出的宇信公司无线部署的方案要点如下。

（1）室内放置 AP，采用 2.4GHz 和 5GHz 双频方式。

（2）2.4GHz 频段可用 1、6、11 信道，5GHz 频段可用 149、153、157、161 和 165 这 5 个频点。

（3）每个 AP 覆盖半径 8～12m。

（4）并发用户数 n=150 人，每个 AP 双频接入用户数按 40 人进行规划，共需 150÷40 ≈ 4 台 AP。

4．AP 部署。

由于半开放环境可视为无阻碍，在满足容量条件下应尽可能少地部署 AP，以减低同邻频信号的干扰，因此建议实际范围内 AP 数量不超过 3 个。该办公区一共 5 层，各层之间需要考虑信号泄露问题，因此各楼间的信道规划需遵循交叉规划原则。确定 AP 类型、数量和安装位置及安装方式后，会议室将独立放置一台 AP。本任务 AP 部署位置如图 5-2-2 所示。

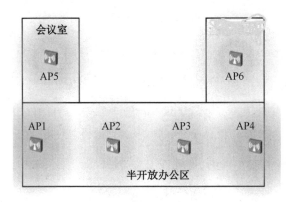

图 5-2-2　AP 部署位置

同时，确定 AP 的取电方式为 PoE 供电。

 办公区楼层平面图

在设计 AP 部署位置时，应根据项目规模及企业办公环境复杂度来确定是否需要办公或楼层平面图。如果客户没有平面图，则可以现场绘制草图，但尺寸和比例要准确。

如果企业需要效果图，则可以使用网规软件或华为云网规导入平面图并进行设计，最终为企业提供网络规划报告和无线覆盖效果图。

 知识补给

当现场测试的覆盖场强大于-65dBm时，1个放装型AP可以覆盖3个小房间（包括左右相邻的两个房间）。若该场景下走廊两侧均有房间，则建议将两侧房间内的AP按照"W"型进行部署。此场景不建议用在走廊中部署放装型AP来覆盖房间的方案中。

对于一般的室内半开放区域，建议AP的间距为20m左右。

确定AP部署位置时，要避开可能产生射频信号的其他电子干扰源，如微波炉、无绳电话、监视器、无线摄像头、红外传感器、荧光灯镇流器等。

【提示：dBm是一个表示功率绝对值的单位，dBm越大，则表示周围信号越好，如-69dBm的信号要比-99dBm的信号好。】

 任务拓展

当企业发展较快，希望所选用的AP型号能够满足近三年内接入数量和带宽的要求时，应该如何对AP选型进行调整呢？

 小贴士

1. 室外AP和室外天线必须采取防雷保护措施，应部署在避雷针45°保护区域内，并就近接地，以免因遭受雷击而造成设备损坏和业务中断。
2. 对于所有室外设备的电缆线、电源线、网线和光纤接头处都必须进行防水处理。

小技巧

华为无线控制器支持对所管理的AP自动规划信道，可以手动配置，也可以开启信道和功率自动调优。

华为云网规工具，可免安装，使用UNIPORTAL账号直接登录即可使用，可对PDF、JPG、PNG、BMP等格式图纸中的障碍物自动识别，无须手动下载新版本、申请license。

5.3 配置AP上线

配置AP上线

> **任务情景**

宇信公司的无线覆盖规划和设备选型已经完成，项目进入实施阶段。按照规划，应在公司的办事大厅部署两个WLAN，一个供公司员工（Stuff）接入，另一个供访客（Guest）接入。

> 任务分析

> 配置无线AP上线；
> 配置无线域模板；
> 配置WLAN的基本参数；
> 查看无线用户接入情况。

> 实施准备

1. HUAWEIS5720交换机1台；
2. HUAWEIAD9430DN管理器1台；
3. HUAWEIAC6005-8-PWR无线接入控制器1台；
4. HUAWEIAP6050DN 1台；
5. 智能手机1部；
6. 直通双绞线2根；
7. Console通信线缆1根；
8. Secure CRT仿真软件1套。

> 实施步骤

1. 按照拓扑图连接交换机与无线设备，其拓扑连接示意图如图5-3-1所示。

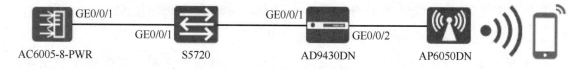

图5-3-1　拓扑连接示意图

2. 在交换机上创建VLAN100、VLAN101、VLAN102，其中VLAN100用于管理流量，VLAN101和VLAN102分别用于Stuff和Guest两个WLAN的数据传输。

```
<HUAWEI> system-view   //进入系统视图
[HUAWEI] sysname Switch  //修改交换机名称
[Switch] vlan batch 100 101 102     //创建VLAN100、VLAN101、VLAN102
[Switch] interface range g0/0/1 to g0/0/2  //进入接口视图
[Switch-port-group] port link-type trunk    //开启中继模式
[Switch-port-group] port trunk allow-pass vlan 100 101 102
//允许VLAN101和VLAN102数据流通过
[Switch-port-group] quit  //返回接口视图
```

3. 在无线AC上创建三个VLAN，其中VLAN100用于无线AP管理流量通过，VLAN101和VLAN102分别用于Stuff和Guest的WLAN数据流。

```
<AC6005> system-view     //进入系统视图
[AC6605] sysname AC       //命名无线AC
```

```
[AC] vlan batch 100 101 102        //创建VLAN100、VLAN101、VLAN102
[AC] interface gigabitethernet 0/0/1   //进入AC与AP互连接口
[AC-GigabitEthernet0/0/1] port link-type trunk      //开启接口骨干模式
[AC-GigabitEthernet0/0/1] port trunk pvid vlan 100    //AP管理流属于VLAN100
[AC-GigabitEthernet0/0/1] port trunk allow-pass vlan 100 101 102 //允许VLAN100和VLAN102
```
数据流通过
```
[AC-GigabitEthernet0/0/1] quit    //返回接口视图
```

4. 开启 AC 的 DHCP 功能，定义三个地址池，分别属于 VLAN100、VLAN101、VLAN102，为无线 AP、Stuff 和 Guest 提供动态地址分配。

```
[AC] dhcp enable       //开启DHCP功能
[AC] interface vlanif 100   //进入VLAN100接口，定义动态地址池
[AC-Vlanif100] ip address 10.100.100.1 24    //网关为10.100.100.1/24
[AC-Vlanif100] dhcp select interface         //本接口作为VLAN100设备的网关
[AC-Vlanif100] quit                          //返回接口视图

[AC] interface vlanif 101    //进入VLAN101接口，定义动态地址池
[AC-Vlanif101] ip address 10.100.101.1 24  //网关为10.100.101.1/24
[AC-Vlanif101] dhcp select interface  //本接口作为VLAN101的网关，分配动态IP地址
[AC-Vlanif101] quit     // 返回接口视图

[AC] interface vlanif 102    //进入VLAN102接口，定义动态地址池
[AC-Vlanif101] ip address 10.100.102.1 24  //网关为10.100.102.1/24
[AC-Vlanif101] dhcp select interface  //本接口作为VLAN102的网关，分配动态IP地址
[AC-Vlanif101] quit     // 返回接口视图
```

5. 开启 AP 电源，执行 display ip pool interface Vlanif100 used 命令，查看 DHCP 获取情况。

```
[AC]display ip pool interface Vlanif100 used
  Pool-name         : Vlanif100
  Pool-No           : 0
  Lease             : 1 Days 0 Hours 0 Minutes
  Domain-name       : -
  DNS-server0       : -
  NBNS-server0      : -
  Netbios-type      : -
  Position          : Interface        Status          : Unlocked
  Gateway-0         : -
  Network           : 10.100.100.0
  Mask              : 255.255.255.0
  Logging           : Disable
  Conflicted address recycle interval: -
  Address Statistic: Total    :254      Used        :1
                     Idle     :253      Expired     :0
                     Conflict :0        Disabled    :0
  --------------------------------------------------------------------
  Network section
      Start          End          Total   Used Idle(Expired) Conflict Disabled
  --------------------------------------------------------------------
    10.100.100.1  10.100.100.254   254     1     253(0)         0        0
  --------------------------------------------------------------------
  Client-ID format as follows:
```

```
         DHCP    : mac-address              PPPoE   : mac-address
         IPSec   : user-id/portnumber/vrf   PPP     : interface index
         L2TP    : cpu-slot/session-id      SSL-VPN : user-id/session-id
         ----------------------------------------------------------------
         Index      IP              Client-ID          Type    Left    Status
         ----------------------------------------------------------------
         34      10.100.100.35    XXXX-XXXX-XXXX       DHCP    86385   Used
         ----------------------------------------------------------------
```
#提示：XXXX-XXXX-XXXX表示本实验中AP的MAC地址

6. 此处可以看到，AP 已获取 10.100.100.35 地址，租期为 86385 秒。

7. 配置 AC 与 AP 的通信 VLAN 及认证方式。

```
[AC] capwap source interface vlanif 100    //配置AC的源接口
#提示：capwap为无线AC与AP通信协议，此命令指定此协议数据从VLAN100通过
[AC] wlan              //进入无线配置视图
[AC-wlan-view] ap auth-mode mac-auth       //配置AC与AP互连的认证方式为MAC地址认证
[AC-wlan-view]ap-id 0 ap-mac xxx-xxxx-xxxx    //写入当前AP的MAC地址，以便AC获知
#提示：AP的MAC地址贴在AP背面的出厂标记上，也可在步骤5中查看IP租用情况时查看并记录
[AC-wlan-ap-0] ap-name AP1    //给AP命名为AP1
Warning: This operation may cause AP reset. Continue? [Y/N] y   //输入Y以确认
#提示：系统提示此操作会引起AP重启，需要再次确认这个操作
[AC-wlan-view]quit    //退出无线配置视图
#提示：AP上电后，当执行命令display ap all后查看到AP的State字段为nor时，表示AP正常上线
[AC]display ap all
Info: This operation may take a few seconds. Please wait for a moment.done.
Total AP information:
nor : normal         [1]
----------------------------------------------------------------
ID  MAC         Name    Group    IP           Type       State  STA Uptime
0   00e0-fcc7-6ac0  AP01  default 10.100.100.117 AP6050DN   nor    0   12S
----------------------------------------------------------------
Total: 1
```
#提示：nor为normal的简写，表示AP状态正常。若此处显示idel，即表示AP处于空闲状态，未进入正常工作状态，可间隔1分钟左右再次查看。如果AP连续处于idel状态，则需要检查无线AC与无线AP的专用通信通道（CAPWAP）是否配置正确

小贴士

为避免出现不法设备拦截信号的问题，无线 AP 获取地址后，需要由 AC 进行认证才可连接。本任务中采取 MAC 地址认证方式，需要在 AC 上建立一个授权的 MAC 地址库，把已知的 AP 地址预先告知 AC。

8. 配置 WLAN 域管理模板，在域管理模板配置 AC 所处国家的国家代码（如中国的国家代码为 cn），并在 AP 组引用域管理模板。

```
[AC] wlan
[AC-wlan-view] regulatory-domain-profile name test    //创建域管理模板，命名为test
[AC-wlan-regulate-domain- test] country-code cn    //国家代码为cn
[AC-wlan-regulate-domain- test] quit
[AC-wlan-view] ap-group name default    //进入默认的AP组
```

```
[AC-wlan-ap-group-ap- default] regulatory-domain-profile test   //调用test模板
Warning: Modifying the country code will clear channel, power and antenna gain configurations of the radio and reset the AP. Continue?[Y/N]:y
[AC-wlan-ap-group-ap- default] quit
```
#提示：Console用户界面已配置完成，用户可以通过Console接口使用密码登录设备，实现本地维护

9．创建名为 wlan-net 的安全模板，并配置安全策略。
```
[AC-wlan-view] security-profile name wlan-net      //创建安全模板，命名为wlan-net
[AC-wlan-sec-prof-wlan-net] security wpa-wpa2 psk pass-phrase a1234567 aes
//指定安全模式为wpa2版本,使用预定义的密码a1234567进行认证
[AC-wlan-sec-prof-wlan-net] quit
```

小贴士

上述例子中，WPA-WPA2+PSK+AES 的安全策略密码为 a1234567，实际配置中请根据实际情况，配置符合要求的安全策略。

10．创建名为 ssid 的 SSID 模板，配置 SSID 名称为 Stuff。
```
[AC-wlan-view] ssid-profile name ssid      //创建SSID模板，并命名为ssid
[AC-wlan-ssid-prof-ssid] ssid Stuff        //供员工接入的SSID
[AC-wlan-ssid-prof-ssid] quit
```

11．创建名为 wlan-net 的 VAP 模板，配置业务数据转发模式为隧道转发模式，配置数据 VLAN 为 VLAN101，并调用安全模板和 SSID 模板。
```
[AC-wlan-view] vap-profile name wlan-net1            //创建VAP模板，并命名为wlan-net
[AC-wlan-vap-prof-wlan-net1] forward-mode tunnel      //选择隧道转发模式
[AC-wlan-vap-prof-wlan-net1] service-vlan vlan-id 101 //绑定用户VLAN为VLAN101
[AC-wlan-vap-prof-wlan-net1] security-profile wlan-net //调用安全模板
[AC-wlan-vap-prof-wlan-net1] ssid-profile ssid         //调用SSID模板
[AC-wlan-vap-prof-wlan-net1] quit
```
#提示：此处创建的VAP模板是将要配置的无线局域网（WLAN）所需的参数的预设值，包括绑定的数据VLAN、WLAN的安全模板、WLAN的SSID模板。本步骤中VAP模板是供Stuff使用的Stuff网络

12．创建名为 ssid2 的 SSID 模板，配置 SSID 名称为 Guest。
```
[AC-wlan-view] ssid-profile name ssid2       //创建SSID模板，命名为ssid2
[AC-wlan-ssid-prof-ssid2] ssid Guest         //配置SSID,并命名为Guest
[AC-wlan-ssid-prof-ssid2] quit
```

13．创建名为 wlan-net2 的 VAP 模板，配置业务数据转发模式为隧道转发模式，配置数据 VLAN 为 VLAN102，并调用安全模板和 SSID 模板。
```
[AC-wlan-view] vap-profile name wlan-net2           //创建VAP模板，并命名为wlan-net2
[AC-wlan-vap-prof-wlan-net2] forward-mode tunnel     //隧道转发模式
[AC-wlan-vap-prof-wlan-net2] service-vlan vlan-id 102 //绑定用户VLAN为VLAN102
[AC-wlan-vap-prof-wlan-net2] ssid-profile ssid2       //调用SSID模板
[AC-wlan-vap-prof-wlan-net2] quit
```
#提示：此处创建的VAP模板是供Guest使用的Guest无线局域网所需的预设参数集，以便在下一步骤中供AP的射频口调用并发射信号

14．配置 AP 组引用 VAP 模板，AP 上射频 0 和射频 1 都使用 VAP 模板 wlan-net 的配置。
```
[AC-wlan-view] ap-group name default
[AC-wlan-ap-group-ap-default] vap-profile wlan-net1 wlan 1 radio 0
```

```
                                               //射频0
[AC-wlan-ap-group-ap-default] vap-profile wlan-net1 wlan 1 radio 1
                                               //射频1
[AC-wlan-ap-group-ap-default] quit
[AC-wlan-view] ap-group name default
[AC-wlan-ap-group-ap-default]vap-profile wlan-net2 wlan 2 radio 0
                                               //射频0
[AC-wlan-ap-group-ap-default]vap-profile wlan-net2 wlan 2 radio 1
                                               //射频1
[AC-wlan-ap-group-ap-default] quit
```

15. 验证配置结果。WLAN业务配置会自动下发给用户,配置完成后,通过执行命令display vap ssid <ssid 名>查看相关信息,当 Status 项显示为 ON 时,表示 AP 对应的射频上的 VAP 已创建成功。

```
[AC-wlan-view] display vap ssid Stuff    // 查看SSID信号状态,配置操作者以自己的名字命名此无线局域网的SSID
WID : WLAN ID
--------------------------------------------------------------------
AP ID AP name   RfID WID  BSSID          Status  Auth type    STA SSID
--------------------------------------------------------------------
0    AP1   0    1   00e0-fcc7-6ac0 ON   WPA/WPA2-PSK   0  Stuff
0    AP1   1    1   00e0-fcc7-6ac0 ON   WPA/WPA2-PSK   0  Stuff
Total: 2
#提示:用户搜索到名为Stuff的无线网络,输入密码a1234567并正常连接无线网络后,在AC上执行display station ssid Stuff命令,可以查看用户是否已经接入无线网络Stuff中
```

16. 查看连接到 WLAN 的无线用户信息。

```
[AC-wlan-view] display station ssidStuff    // 查看连接到Stuff网络下的用户
Rf/WLAN: Radio ID/WLAN ID
Rx/Tx: link receive rate/link transmit rate(Mbps)
STA MAC     AP ID Ap name    Rf/WLAN  Band  Type  Rx/Tx  RSSI  VLAN  IP address
--------------------------------------------------------------------
9487-e01d-4f86 0 AP1      1/1     5G    11n   46/59  -68   101   10.100.101.104
--------------------------------------------------------------------
Total: 1 2.4G: 0 5G: 1
```

小贴士

查看无线用户设备（STA）信息，将依次显示该设备的 MAC 地址、所连 AP 的 ID 及名字、射频 ID、WLAN ID、信号频段、802.11 协议类型、收/发速率、信号强度、VLAN ID、IP 地址等，Total 提示后显示的是汇总信息。

小技巧

在 AC 控制器上，默认情况下，AP 认证模式为 MAC 地址认证。为了使 AP 批量上线，可使用 ap auth-mode no-auth 设置不对 AP 认证，等待要开通的 AP 上线以后再调整为 MAC 地址认证。

知识补给

FIT AP（瘦 AP）控制架构 + 无线控制器（AC）对设备的功能进行了重新划分，其中无线控制器负责无线网络的接入控制、转发和统计、AP 的配置监控、漫游管理、AP 的网管代理、安全控制；FIT AP 负责 802.11 报文的加解密、按照 802.11 协议进行物理层功能、接收无线控制器的管理、RF 空口的统计等简单功能的设置。

1. FIT AP+AC 组网方式。

（1）FIT AP 与 AC 之间的网络架构可分为二层组网和三层组网。

在二层网络连接模式中，FIT AP 和 AC 属于同一个二层广播域，FIT AP 和 AC 之间通过二层交换机互连，如图 5-3-2 所示。

当 FIT AP 与 AC 之间的网络为三层网络时，WLAN 组网为三层组网，如图 5-3-3 所示。在实际组网中，一台 AC 可以连接几十台甚至几百台 FIT AP，组网一般比较复杂。在企业网络中，FIT AP 可以安放在办公室、会议室、会客间等场所，而 AC 可以安放在公司机房，这样，FIT AP 和 AC 之间的网络就是比较复杂的三层网络。在大型组网中一般采用三层组网。

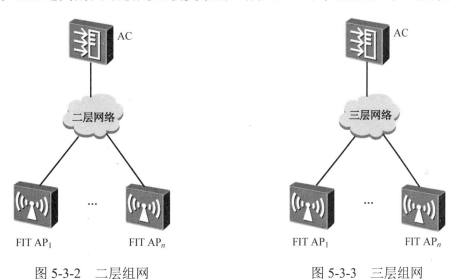

图 5-3-2　二层组网　　　　　　　　图 5-3-3　三层组网

（2）根据 AC 在网络中的位置可分为直连式组网和旁挂式组网。

直连式组网是将 FIT AP、AC 与上层网络串联在一起，所有数据必须通过 AC 访问上层网络，如图 5-3-4 所示。

采用这种组网方式时，对 AC 的吞吐量及处理数据量的要求比较高，否则 AC 会是整个无线网络带宽的瓶颈。此种组网方式的组网架构清晰，组网实施起来比较简单。

旁挂式组网是指将 AC 旁挂在 FIT AP 与上层网络的直连网络中，不再直接连接 FIT AP，FIT AP 的业务数据可以不经过 AC 而直接到达上层网络，如图 5-3-5 所示。

实际组网时，大部分网络并不是早期就规划好的，无线网络的覆盖架构是后期在原有网络的基础上扩展而来的。采用旁挂式组网比较容易进行扩展，只需将 AC 旁挂在现有网络中，如旁挂在汇聚交换机上,就可以对终端 FIT AP 进行管理。所以此种组网方式的使用率比较高。

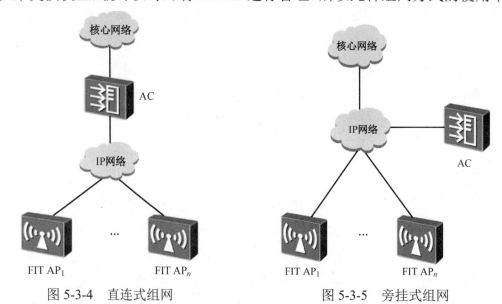

图 5-3-4　直连式组网　　　　　　　图 5-3-5　旁挂式组网

在旁挂式组网中,AC 只承载对 FIT AP 的管理功能,管理流在 CAPWAP 数据隧道中传输。数据业务流可以通过 CAPWAP 数据隧道经过 AC 转发,也可以不经过 AC 就直接转发,后者的无线用户业务流经汇聚交换机传输至上层网络。

2. WLAN 配置流程。

本任务中 WLAN 架构采用二层直连式组网,CAPWAP 协议数据(AC 与 FIT AP 管理信息)从 VLAN100 通过,STA 数据(无线用户设备)分别从 VLAN101、VLAN102 通过。DHCP 服务由 AC 提供。WLAN 数据规划如表 5-3-1 所示。

表 5-3-1　WLAN 数据规划

配置项	数据	
DHCP 服务器	AC 作为 DHCP 服务器,为用户和 AP 分配 IP 地址	
AP 的 IP 地址池	10.100.100.2～10.1.100.254/24	
STA 的 IP 地址池	VLAN101：10.100.101.2～10.1.101.254/24 VLAN102：10.100.102.2～10.1.102.254/24	
AC 的源接口 IP 地址	10.100.100.1/24	
AP 组	名称：default(默认组) 引用模板：VAP 模板、域管理模板	
域管理模板	名称：test；国家代码：cn	
SSID 模板	SSID 名称：ssid	SSID2 名称：ssid2
安全模板	名称：wlan-net；WPA-WPA2+PSK+AES 加密方式	
VAP 模板	名称：wlan-net 转发模式：隧道转发 业务 VLAN：101 引用模板：SSID 模板 ssid、安全模板 wlan-net	名称：wlan-net2 转发模式：隧道转发 业务 VLAN：102 引用模板：SSID 模板 ssid2

AC 配置过程包括四个部分，详细配置流程如图 5-3-6 所示。

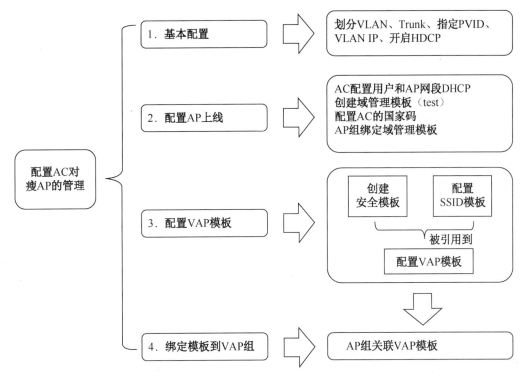

图 5-3-6 WLAN 配置流程

5.4 测试 WLAN 性能

> 任务情景

经过现场勘察、前期规划和设备配置，宇信公司的无线局域网覆盖项目接近尾声。在验收之前，小李在高工程师的指导下，对部署的无线网络进行性能测试，以检验办公区域内是否能够满足无线终端接入、漫游、移动办公及上网互通等需求。

> 任务分析

> 学会测试 WLAN 覆盖范围；
> 学会测试 WLAN 信号强度；
> 学会测试无线用户的带宽限制；
> 学会测试基于 SSID 的带宽限制。

> 实施准备

1．笔记本电脑 1 台，预装无线测速软件（本任务中使用 WirelessMon Professional）；
2．智能手机 1 部，支持 2.4G/5G 接入；
3．AC6005 和 AP 之间用线缆连接并配置 WLAN；

4. Console 通信线缆 1 根。

➢ 实施步骤

1. 为建立正确的测试环境，需首先检查此 WLAN 中的关联设备的工作是否正常，并在 AC 上配置无线业务，设置 SSID 为 Guest，接入时采用"不验证"方式。

2. 在距离 AP 5m、10m、15m、30m……处，分别用笔记本电脑和智能手机寻找并关联 SSID 为"Guest"的 WLAN。关联成功后，查看 STA 是否可以获取到预设的 IP 地址，判断是否成功关联到 STA，并判断是否成功关联到 Guest 的网络，并且正确获取 IP 地址，如图 5-4-1 所示。

图 5-4-1　正确连接 Guest 网络

3. 打开笔记本电脑上的测试软件，测试本次关联的 WLAN 的信号强度，以及所用信道、传输速率等信息，如图 5-4-2 所示。

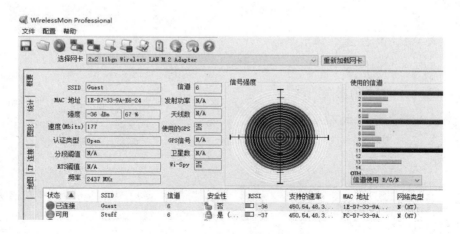

图 5-4-2　WLAN 的信号强度

通过测速可以看出，本任务部署的“Guest”和“Stuff”的信号强度分别为-36dBm、-37dBm，工作性能良好，终端设备易接入。

小贴士

WLAN 信号强度的范围一般为-30～-120dBm，正常信号强度应该是-40～-85dBm。本任务的信号强度为-36dBm，已经很强了，基本上没有什么衰减，是非常好的网络连接。小于-90 dBm 就很差了，几乎无法连接。

4．分别从不同方向，重复进行步骤 2 的测试，记录能够成功连接 Guest 网络的安全距离，即为该 WLAN 的覆盖范围。

5．在 AC 上配置基于用户的带宽限制功能，并限制用户的上行流量带宽是 1MB/s，下行流量带宽是 2MB/s。

```
    <HUAWEI AC6005> system-view
    [HUAWEI AC6005] wlan
    [HUAWEI AC6005-wlan-view] traffic-profile name default   //配置流量脚本，命名为default
    [HUAWEI AC6005-wlan-traffic-prof-default]rate-limit vap up 10240
//限制上行流量带宽为10MB/s
    [HUAWEI AC6005-wlan-traffic-prof-default]rate-limit vap down 30720
//限制下行流量带宽为30MB/s
```

6．使用便携式计算机，打开 360 安全卫士软件，在该软件中选择“宽带测试器”选项以进行测试，便携式计算机和智能手机的上行流量均被限制在 5MB/s 左右，下行流量均被限制在 30MB/s 左右，则本地无线用户的带宽限速成功，测试页面如图 5-4-3 所示。

IP	223.88.46.163	测试结果
来自	河南省郑州市 移动	上传速度：5.38 MB/秒（上传带宽：43.01 Mb）
操作系统	Windows 10	
浏览器	Google Chrome	下载速度：5.18 MB/秒（下载带宽：41.44 Mb）
测试时间	2020-02-14 16:59:32	
预估您的宽带为 41.44 MB		

图 5-4-3　测试页面

小贴士

如图 5-4-3 所示中的 IP 地址是经过出口路由器进行 NAT 转换的，所以显示为公有 IP 地址。

7．为保障网络质量，可在 AC 上配置接入用户数。为了测试方便，此处配置最大接入用户数为 2。因此需要在 AC 上做如下配置：

```
    <HUAWEI AC6005> system-view
    [HUAWEI AC6005] authentication-profile name default
```

```
[HUAWEI AC6005-authentication-profile-default]authentication wlan-max-user 2
```

8. 使用笔记本电脑（STA1）和智能手机（STA2）均能正常关联 Guess 并获取地址，且能 ping 通网关，在 AC 上可以查看到二者的接入信息；此时再使用第三台无线 STA 设备，则无法接入网络，在 AC 上只能查看到 STA1~STA2 的接入信息。

任务拓展

为进一步控制 WLAN 的安全，可以在无线 AP 上设置黑名单或白名单，以限制 STA 的接入数量。

1. 创建名为"sta-whitelist"的 STA 白名单模板，将 STA1 的 MAC 地址加入白名单。

```
[AC-wlan-view] sta-whitelist-profile name sta-whitelist    //设置白名单，命名为sta-whitelist
[AC-wlan-whitelist-prof-sta-whitelist] sta-mac 0011-2233-4455
//把STA1的MAC地址加入白名单
[AC-wlan-whitelist-prof-sta-whitelist] quit    //返回白名单配置模式
# 在VAP模板wlan-net中引用STA白名单模板，使白名单在VAP范围内有效
[AC-wlan-view] vap-profile name wlan-net    //进入名称为wlan-net的VAP配置模板
[AC-wlan-vap-prof-wlan-net] sta-access-mode whitelist sta-whitelist
//引用名称为sta-whitelist的白名单
[AC-wlan-vap-prof-wlan-net] quit    //返回VAP配置模板编辑模式
```

2. 创建名为"sta-blacklist"的 STA 黑名单模板，将 STA 的 MAC 地址加入黑名单。

```
[AC-wlan-view] sta-blacklist-profile name sta-blacklist    //设置黑名单，命名为sta-blacklist
[AC-wlan-blacklist-prof-sta-blacklist] sta-mac 0011-2233-4455    //把STA1的MAC地址加入黑名单
[AC-wlan-blacklist-prof-sta-blacklist] quit    //返回黑名单配置模式
#创建名为wlan-system的AP系统模板，并引用STA黑名单模板，使黑名单在AP范围内有效
[AC-wlan-view] ap-system-profile name wlan-system    //创建名为wlan-system的AP系统模板
[AC-wlan-ap-system-prof-wlan-system] sta-access-mode blacklist sta-blacklist
//引用名称为sta-blacklist的黑名单
[AC-wlan-ap-system-prof-wlan-system] quit    //返回AP系统模板编辑模式
```

3. 接入测试。

（1）确保所有设备工作正常，断开已有无线连接。

在 AC 上进行步骤 1 中开启白名单的配置，用 STA1、STA2 同时关联 Guest 无线网络，STA1 能成功关联到 Guest，而 STA2 则无法关联。

（2）断开已有连接。在 AC 上进行步骤 2 中开启黑名单的配置，用 STA1、STA2 同时关联 Guest 无线网络，STA2 能成功关联到 Guest，而 STA1 则无法关联。

小贴士

AC 只能选择黑名单或白名单中的一种限制方式，二者不能同时选用。

5.5 认识防火墙

➢ 任务情景

宇信公司为企业网配置了华为防火墙,以作为公司的出口网关设备。在接入防火墙之前,小李需要首先熟悉防火墙的外观、接口和指示灯,并进行初始配置。

➢ 任务分析

- ➢ 认识防火墙硬件接口及功能;
- ➢ 了解防火墙指示灯状态信息;
- ➢ 掌握防火墙登录方法。

➢ 实施准备

1. HUAWEIUSG6510 防火墙 1 台;
2. PC 1 台;
3. UTP 通信线缆 1 根。

➢ 实施步骤

1. 将 UTP 通信线缆的一端插入 PC 的网口中,再将另一端插入防火墙设备的 GE0/0/0(MGMT)中,如图 5-5-1 所示。

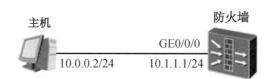

图 5-5-1 线缆连接示意图

🐟 小贴士

在连线时使用普通的百兆或千兆网卡都可以,目前设备和 PC 的接口都支持自适应,所以无论选择交叉线缆还是直通线缆都可以。

2. 验证管理员登录防火墙(FireWall)。
（1）配置管理员 PC 的 IP 地址为 192.168.0.2/24。
（2）在 PC 中打开网络浏览器,访问设备的 IP 地址"https://192.168.0.1:8443"。
（3）在登录界面中输入管理员的用户名"admin"和密码"Admin@123",单击"登录"按钮,进入防火墙配置界面,说明管理配置成功,如图 5-5-2 所示。防火墙配置界面功能区域划分如图 5-5-3 所示。

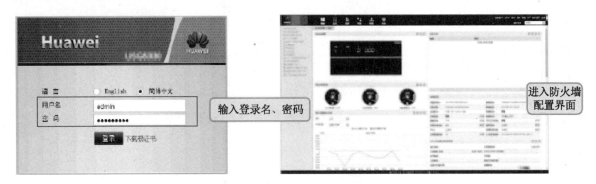

图 5-5-2　登录防火墙界面

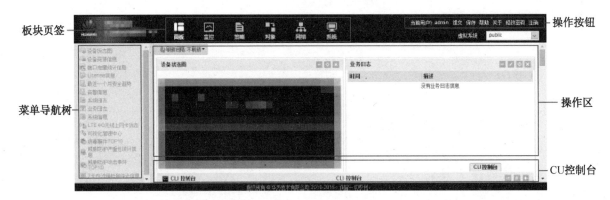

图 5-5-3　防火墙配置界面功能区域划分

小贴士

输入地址并登录后，浏览器会给出证书不安全的提示，此时可以选择继续浏览。在登录界面单击"下载根证书"按钮，并导入管理员 PC 的浏览器，再次登录时就不会出现告警提示了，如图 5-5-4 所示。

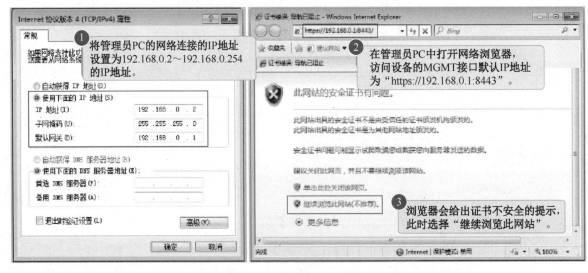

图 5-5-4　登录防火墙界面

3. 创建管理员并指定认证类型、管理员角色及信任主机。

（1）在登录防火墙界面，依次选择"系统"→"管理员"→"管理员"，如图 5-5-5 所示。

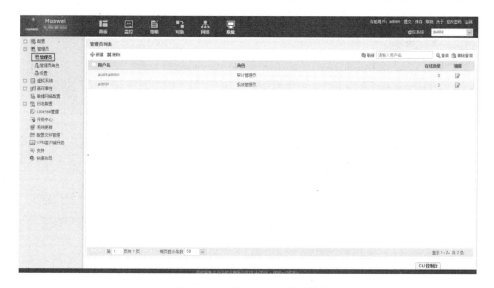

图 5-5-5　登录防火墙界面

（2）单击"新建"按钮，参数配置如图 5-5-6 所示。

图 5-5-6　参数配置

4. 配置 Web 服务的超时时间。

（1）依次选择"系统"→"管理员"→"设置"，配置 Web 服务的超时时间如图 5-5-7 所示。

图 5-5-7　配置 Web 服务的超时时间

（2）在"Web 服务超时时间"文本框中输入"5"。此处数据"5"代表 5 分钟。

（3）单击"应用"按钮，使以上配置生效。

任务总结与思考

对一台新出厂的防火墙设备进行业务配置时，受限于本地登录的配置信息。本任务重点讲述设备首次登录时采取 Web 登录的具体操作方法。

思考以下两个问题。

1. 设备首次登录方式除 Web 登录方式外，还有没有其他的登录方式？

2. 使用防火墙的 Web 管理方式时，可以登录 GE0/0/0 的默认 IP 地址进行访问。如果该防火墙的其他接口也添加了 IP 地址，则能否通过登录其他接口 Web 进入防火墙？

知识补给

防火墙的硬件知识

通过对以上任务的学习，大家已经了解了防火墙的登录流程和登录方法，接下来了解一下防火墙的硬件知识，重点关注防火墙前面板指示灯的用途，其外观如图 5-5-8 所示，指示灯功能介绍如表 5-5-1 所示。

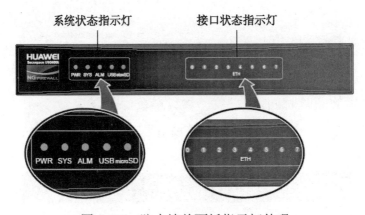

图 5-5-8　防火墙前面板指示灯外观

表 5-5-1　防火墙前面板指示灯功能介绍

指示灯名称	功能描述
接口状态指示灯 0～7（绿色）	常亮：链路已经连通 每秒闪 8 次（8Hz）：有数据收发 常灭：链路没有连通
系统状态指示灯	常亮：表示系统处于工作状态
PWR 指示灯（绿色）	常亮：电源工作正常 常灭：电源故障或没有连接

续表

指示灯名称	功能描述
SYS 指示灯（绿色）	常亮：系统处于上电加载或复位启动状态 每 2 秒闪 1 次：系统处于正常运行状态 每秒闪 2 次：系统处于启动中 每秒闪 8 次：系统软件或配置文件正在升级 常灭：系统故障
ALM 指示灯（红色）	常亮：系统运行出现故障，如整机上电启检、电压、温度检测出现异常 常灭：系统运行正常
USB 指示灯（绿色）	常亮：USB2.0 接口已经连接 常灭：USB2.0 接口没有连接
micro SD 指示灯（绿色）	常亮：micro SD 卡在位 常灭：micro SD 卡不在位

相对于前面板，防火墙背板上的接口相对较少，其外观如图 5-5-9 所示，功能介绍如表 5-5-2 所示。

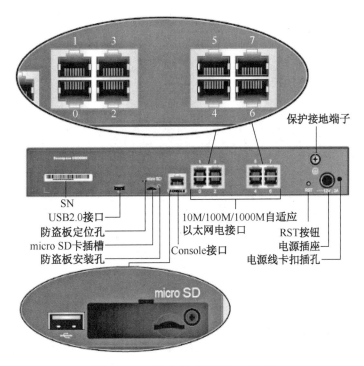

图 5-5-9　防火墙背板接口外观

表 5-5-2　防火墙背板接口功能介绍

接口名称	功能描述
SN	标识设备的数字序列号，申请 license 文件时需要提供设备的 SN 信息
USB2.0 接口	插接 U 盘，通过 U 盘升级设备的系统软件等
micro SD 卡插槽	插接 micro SD 卡，为用户提供实时记录日志和报表。micro SD 卡为可选配置

续表

接口名称	功能描述
Console 接口	Console 接口用于连接控制台，实现本地配置功能。通过 Console 接口配置线缆连接 Console 接口到 PC 的串口（com），使用串口终端软件访问设备的命令配置界面，输入相应命令后对设备进行配置、管理和维护
0～7（RJ45）	8 个 10M/100M/1000M 自适应以太网电接口，接口编号为 GE0/0/0～0/0/7。GE0/0/0 是带内管理口，默认 IP 地址为 192.168.0.1。可使用网线连接该接口和 PC，并通过 Telnet、sTelnet、Web 等方式登录设备，对设备进行配置、管理和维护
RST 按钮	当设备正常运行时，按下 RST 按钮将重新启动设备。建议按 RST 按钮前保存当前配置。该按钮还可用于一键恢复默认配置，即在设备上电前先按住 RST 按钮 3～5s，当前面板上的 SYS 指示灯闪烁时，松开 RST 按钮，设备会使用默认配置启动
电源插座	用于连接电源适配器的 4PIN 插头端
电源线卡和插孔	用于安装电源线卡扣。该卡扣用来绑定电源线，防止电源线松脱
保护接地端子	用于连接保护地线的 M4.0T 端子，将保护底线连接到机柜、工作台、墙体的接地点或机房中的接地排上

任务拓展

在本任务的情景与拓扑结构都不变的情况下，请尝试通过其他接口，采用 Web 方式登录防火墙设备，组网连接如图 5-5-10 所示。

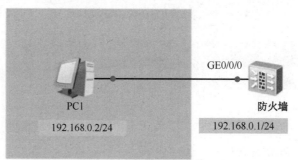

图 5-5-10　组网连接

登录说明如下。

1．通过默认 GE0/0/0 管理接口登录防火墙 Web 界面。

2．选择"网络"→"接口"。

3．单击 GE1/0/3 接口右侧的 ☑（其他接口配置方法相同），按如图 5-5-11 所示方式配置参数。

安全区域	trust 区域
连接类型	静态 IP
IP 地址	10.3.0.1/24
启动访问管理	HTTPS

图 5-5-11　配置参数

4. 单击"确定"按钮。
5. 配置管理员 PC 的 IP 地址为 10.3.0.10/24。
6. 在 PC 中打开浏览器，访问设备的 IP 地址"https://10.3.0.1:8443"。
7. 在登录界面中输入管理员的用户名和密码，登录设备。

小贴士

Web 登录和配置方式对于管理员来说要比使用 CLI 命令行方式简单，可使学习成本有效降低，所以目前在防火墙设备上普遍采用 Web 登录方式。但是防火墙和路由器一样也可以支持 Console 接口登录，通过命令行对防火墙配置可以达到 Web 配置相同的效果，在 Web 界面上配置的设备硬件功能会自动转成命令行方式显示在命令行界面，比如在图 5-5-11 中配置了 GE1/0/3 接口的 IP 地址，利用底层命令行可以查到如图 5-5-12 所示信息。

```
#
interface GigabitEthernet1/0/3
 ip address 10.3.0.1 255.255.255.0
 service-manage https permit
#
```

图 5-5-12　查看信息

5.6　划分安全区域

➤ 任务情景

宇信公司新部署的办公网络的出口设备为路由器，随着企业规模的扩大，信息安全的需求也日渐凸显。小李决定购买华为防火墙来替代传统的出口设备路由器，以实现更安全的内外网对接。登录防火墙后，首先需要对防火墙进行网络基础配置，将各个业务接口加入安全区域。一般情况下，将连接外网的接口加入安全级别低的安全区域（如 untrust 区域），将连接内网的接口加入安全级别高的安全区域（如 trust 区域），服务器可以加入 dmz 区域。另外，还需要配置设备名称、时钟、接口 IP 地址、默认路由及默认包过滤等。

➤ 任务分析

- ➤ 掌握防火墙的基本配置；
- ➤ 了解防火墙的区域配置方法；
- ➤ 学会把接口加入对应的安全区域。

➢ 实施准备

1. HUAWEIUSG6510 防火墙 1 台；
2. PC 1 台；
3. AR3260 路由器一台；
4. Console 通信线缆 1 根；UTP 通信线缆 2 根。

➢ 实施步骤

1. 按照企业网络出口需求连接网络设备，如图 5-6-1 所示，各设备接口 IP 地址规划如表 5-6-1 所示。

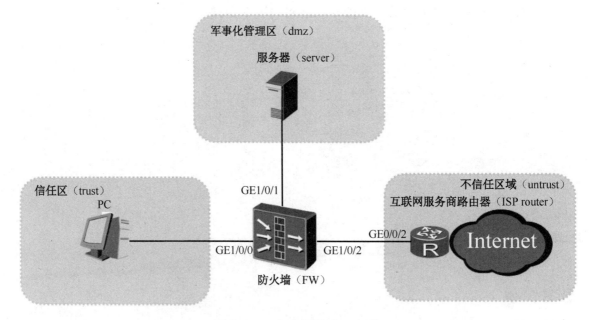

图 5-6-1　企业网络出口拓扑

表 5-6-1　设备接口 IP 地址规划

设备名称	接口	IP 地址	子网掩码
PC	GE0/0/0	192.168.1.2	255.255.255.0
FW	GE1/0/0	192.168.1.1	255.255.255.0
server	GE0/0/0	10.1.1.2	255.255.255.0
FW	GE1/0/1	10.1.1.1	255.255.255.0
ISP router	GE1/0/2	1.1.1.254	255.255.255.0
FW	GE1/0/2	1.1.1.1	255.255.255.0

2. 配置 PC 和服务器的 IP 地址，具体配置参数分别如图 5-6-2 和图 5-6-3 所示。

图 5-6-2　PC 的 IP 地址配置

图 5-6-3　服务器的 IP 地址配置

3．配置防火墙设备名称及接口地址。

配置设备名称。

```
<FW_A>system-view//进入系统视图
[FW] sysname FW_A//路由器启动后默认设备名称为Huawei,此处修改为FW_A
[FW_A] quit//返回用户视图
```

配置时钟，包括当前时间和时区。

```
<FW_A>clock datetime 18:10:45 2019-03-01//设置年月日时分秒
<FW_A>clock timezone BJ add 08:00:00//设置本地时区
```

配置接口的地址。

```
<FW_A>system-view//进入系统视图
[FW_A] interface GigabitEthernet 1/0/0   //进入GE1/0/0接口配置视图
[FW_A-GigabitEthernet1/0/0] ip address 192.168.1.1 24   //配置GE1/0/0的接口地址为192.168.1.1,并设置掩码为24位
[FW_A-GigabitEthernet1/0/0] quit   //返回系统视图
[FW_A] interface GigabitEthernet 1/0/1   //进入GE1/0/1接口配置视图
[FW_A-GigabitEthernet1/0/1] ip address 10.1.1.1 24   //配置GE1/0/1的接口地址为10.1.1.1,并设置子网掩码为24位
[FW_A-GigabitEthernet1/0/1] quit   //返回系统视图
[FW_A] interface GigabitEthernet 1/0/2   //进入GE1/0/2接口配置视图
[FW_A-GigabitEthernet1/0/2] ip address 1.1.1.1 24
//配置GE1/0/1的接口地址为1.1.1.1,并设置子网掩码为24位
[FW_A-GigabitEthernet1/0/2] quit   //返回系统视图
```

4．按照功能需要，将各个业务接口加入安全区域。

```
[FW_A] firewall zone trust//进入trust区域
[FW_A-zone-trust] add interface GigabitEthernet 1/0/0   //将接口加入trust区域
[FW_A-zone-trust] quit//返回系统视图
[FW_A] firewall zone dmz//进入dmz区域
[FW_A-zone-dmz] add interface GigabitEthernet 1/0/1   //将接口加入trust区域
[FW_A-zone-dmz] quit//返回系统视图
[FW_A] firewall zone untrust   //进入untrust区域
[FW_A-zone-untrust] add interface GigabitEthernet 1/0/2   //将接口加入untrust区域
```

```
[FW_A-zone-untrust] quit  //返回系统视图
```

5. 在防火墙上配置默认路由，允许将数据包转发至互联网。
```
[FW_A] ip route-static 0.0.0.0 0.0.0.0 1.1.1.254
//添加到Internet公网的出口路由
```

6. 打开默认包过滤，保证防火墙能够接入Internet。
```
[FW_A] security-policy//进入安全策略视图
[FW_A-policy-security] default action permit  //默认情况下，防火墙不允许域间互访，此命令用于配置默认安全策略为permit动作
```

7. 配置ISP router，添加从网络服务商返回企业内部网的路由。
```
<Huawei>system-view    //进入系统视图
[Huawei]sysname RTA    //修改路由器名称为RTA
[RTA]interface GigabitEthernet 0/0/0//进入GE0/0/0接口配置视图
[RTA-GigabitEthernet0/0/0]ip address 1.1.1.254 24    //配置GE0/0/0的接口地址为1.1.1.254，并设置掩码为24位
  Dec 12 2018 16:32:24+00:00 RTA %%01IFNET/4/LINK_STATE(l)[0]:The line protocol IP on the interface GigabitEthernet0/0/0has entered the UP state.
#提示：此处系统提示接口GE0/0/0的协议状态为UP，说明接口配置成功
[RTA-GigabitEthernet0/0/0]quit   //返回接口GE0/0/0配置视图
[RTA]ip route-static 192.168.1.0 255.255.255.0 1.1.1.1//添加静态路由，用于指导PC1数据包返回局域网的路径
```

8. 测试PC能否ping通ISP router。
```
C:\Users\ >ping 1.1.1.254
正在 Ping 1.1.1.254具有 32 字节的数据：
来自 1.1.1.254的回复：字节=32 时间=1ms TTL=64
来自 1.1.1.254的回复：字节=32 时间=1ms TTL=64
来自 1.1.1.254的回复：字节=32 时间=1ms TTL=64
来自 1.1.1.254的回复：字节=32 时间=1ms TTL=64
1.1.1.254的 Ping 统计信息：
数据包：已发送 = 4，已接收 = 4，丢失 = 0 (0% 丢失)，
往返行程的估计时间(以毫秒为单位)：
最短 = 1ms，最长 = 1ms，平均 = 1ms
#提示：PC1通过ping命令向IP为1.1.1.254的PC2发送4个ICMP包，收到了4个回复，每个包的字节数为32，生存时间为64，耗费时间为1ms，丢包率为0%，说明二者之间网络互通，状态良好
```

任务总结与思考

本任务讲述了防火墙的基础配置，接口地址、接口加入安全区域的配置方法，实现了部门之间的访问隔离。

思考以下两个问题。

1. 接口加入安全区域的作用是什么？
2. 接口不加入安全区域能转发数据吗？

知识补给

为了在防火墙上区分不同的网络，我们在防火墙上引入了一个重要的概念：安全区域

（Security Zone）。安全区域是一个或多个接口的集合，是防火墙区别于路由器的主要特性。防火墙通过安全区域来划分网络、标识报文流动的"路线"。通常，当报文在不同的安全区域之间流动时，就会受到控制。

防火墙通过接口来连接网络，将接口划分到安全区域后，通过接口就可以把安全区域和网络关联起来。通常说某个安全区域时，就特指该安全区域中接口所连接的网络。同样，通过把接口划分到不同的安全区域中，就可以在防火墙上划分出不同的网络。

在华为防火墙上，一个接口只能加入一个安全区域。

1. 华为防火墙产品上默认已经提供了三个安全区域，分别是 trust、dmz 和 untrust：

（1）trust 区域：该区域内网络的受信任程度高，通常用来定义内部用户所在的网络。

（2）dmz 区域：该区域内网络的受信任程度中等，通常用来定义内部服务器所在的网络。

（3）untrust 区域：该区域代表的是不受信任的网络，通常用来定义 Internet 等不安全的网络。

2. 在网络数量较少、环境简单的场合中，使用系统默认提供的安全区域就可以满足划分网络的需求。在网络数量较多的场合，可以根据需要创建新的安全区域。

3. 假设接口 1 和接口 2 连接的是内部用户，那么就可以把这两个接口划分到 trust 区域中；若接口 3 连接的是内部服务器，则可将它划分到 dmz 区域；若接口 4 连接的是 Internet，则可将它划分到 untrust 区域。

4. 当内部网络中的用户访问 Internet 时，报文在防火墙上的路线为从 trust 区域到 untrust 区域；当 Internet 上的用户访问内部服务器时，报文在防火墙上的路线为从 untrust 区域到 dmz 区域。

5. dmz 起源于军方，是介于严格的军事管制区和松散的公共区域之间的一种部分管制区域。防火墙引用了这一术语，指代一个与内部网络和外部网络分离的安全区域。

6. 防火墙上提供了 Local 区域，代表防火墙本身。凡是由防火墙主动发出的报文均可被认为是从 Local 区域中发出的，凡是需要防火墙响应并处理（而不是转发）的报文均可被认为是由 Local 区域接收的。

7. Local 区域中不能添加任何接口，但防火墙上所有接口本身都隐含地属于 Local 区域。也就是说，报文通过接口去往某个网络时，目的安全区域是该接口所在的安全区域；报文通过接口到达防火墙本身时，目的安全区域是 Local 区域。

在防火墙上用安全区域来表示网络后，如何来判断一个安全区域的受信任程度呢？在华为防火墙上，每一个安全区域都有一个唯一的安全级别，且用 1~100 的数字表示，数字越大，则代表该区域内的网络越可信。对于默认的安全区域，它们的安全级别是固定的。例如，Local 区域的安全级别是 100，trust 区域的安全级别是 85，dmz 区域的安全级别是 50，untrust 区域的安全级别是 5。

任务拓展

防火墙通过划分区域，可以把一个网络分成多个模块，并实现各个模块间的访问控制，那么防火墙是工作在二层还是三层呢？

 思考与实训 5

一、选择题

1. 关于 5G 优先特性的描述，下列说法中正确的是（　　）。
   A．5G 优先特性是指让 5G 的客户端优先发送数据
   B．5G 优先开启以后，只支持 2.4GHz 频段的客户端的网络转输速度会变慢
   C．5G 优先要求 AP 在双频时才可以正常工作
   D．5G 优先要求所有的客户端都要同时支持 2.4GHz 和 5GHz 频段

2. 若要获取更好的覆盖效果，应尽量使 AP 信号能够（　　）穿过墙壁。
   A．垂直   B．平行
   C．45°   D．60°

3. 无线网络性能测试可能包含（　　）。
   A．网络延时测试   B．信号强度测试
   C．传输带宽测试   D．终端漫游测试

4. 无线通信过程中信号强度太弱、错误率较高，无线客户端切换到其他无线 AP 的信道，这个过程称为（　　）。
   A．关联   B．重关联
   C．漫游   D．负载平衡

二、填空题

1. WLAN 无线局域网络设备包含_____、_____和无线接入终端。
2. WLAN 射频有_____、_____两个频段。
3. WLAN 网络中无线接入点工作在 OSI 模型中的_____层。
4. 无线控制器未额外购买 AP 接入授权时，最多能管理_____个 AP。
5. 按照在网络中的位置区分，华为 AC 支持_____、_____两种组网方式。
6. 对 WLAN 的 2.4GHz 频段有影响的电子设备有_____、_____、_____、_____等。
7. 华为的 AP 按照覆盖情况分为_____、_____、_____三类。

三、判断题

1．华为无线控制器可以管理不同厂商的 AP 接入点。（   ）

2．通常瘦 AP 需要 AC 来下发配置和管理，胖 AP 类似支持 WiFi 的路由器，不需要 AC 管理和下发配置。（   ）

3．为了使 WLAN 网络信号覆盖得更好，在用户预算充足情况下，应尽量多地规划一些 AP，两个 AP 之间的间隔距离可以小于 10m。（   ）

4．小李在华为公司的官网上查到某个 AP 的射频参数，其最大接入用户数小于等于 256，这表示在同一时刻用户数少于 256 时，用户同时并发上网不会有问题。（   ）

5．USG 防火墙出厂时设置的默认登录账号是 admin，密码是 Huawei@123。（   ）

6．Web 登录防火墙除可以用 GE0/0/0 接口的出厂默认 IP 登录外，无法用其他接口和地址登录。（   ）

7．防火墙既可以支持串口登录，又可以支持 Web 登录。（   ）

反侵权盗版声明

电子工业出版社依法对本作品享有专有出版权。任何未经权利人书面许可，复制、销售或通过信息网络传播本作品的行为；歪曲、篡改、剽窃本作品的行为，均违反《中华人民共和国著作权法》，其行为人应承担相应的民事责任和行政责任，构成犯罪的，将被依法追究刑事责任。

为了维护市场秩序，保护权利人的合法权益，我社将依法查处和打击侵权盗版的单位和个人。欢迎社会各界人士积极举报侵权盗版行为，本社将奖励举报有功人员，并保证举报人的信息不被泄露。

举报电话：（010）88254396；（010）88258888

传　　真：（010）88254397

E-mail：　dbqq@phei.com.cn

通信地址：北京市万寿路 173 信箱

　　　　　电子工业出版社总编办公室

邮　　编：100036